Tobias Sauter

Non-parametric modelling in Geoscience

Tobias Sauter

Non-parametric modelling in Geoscience

Application, optimization and uncertainty estimation

Südwestdeutscher Verlag für Hochschulschriften

Impressum/Imprint (nur für Deutschland/only for Germany)
Bibliografische Information der Deutschen Nationalbibliothek: Die Deutsche Nationalbibliothek verzeichnet diese Publikation in der Deutschen Nationalbibliografie; detaillierte bibliografische Daten sind im Internet über http://dnb.d-nb.de abrufbar.
Alle in diesem Buch genannten Marken und Produktnamen unterliegen warenzeichen-, marken- oder patentrechtlichem Schutz bzw. sind Warenzeichen oder eingetragene Warenzeichen der jeweiligen Inhaber. Die Wiedergabe von Marken, Produktnamen, Gebrauchsnamen, Handelsnamen, Warenbezeichnungen u.s.w. in diesem Werk berechtigt auch ohne besondere Kennzeichnung nicht zu der Annahme, dass solche Namen im Sinne der Warenzeichen- und Markenschutzgesetzgebung als frei zu betrachten wären und daher von jedermann benutzt werden dürften.

Coverbild: www.ingimage.com

Verlag: Südwestdeutscher Verlag für Hochschulschriften GmbH & Co. KG
Heinrich-Böcking-Str. 6-8, 66121 Saarbrücken, Deutschland
Telefon +49 681 37 20 271-1, Telefax +49 681 37 20 271-0
Email: info@svh-verlag.de

Approved by: Aachen, RWTH Aachen University, Diss., 2011

Herstellung in Deutschland:
Schaltungsdienst Lange o.H.G., Berlin
Books on Demand GmbH, Norderstedt
Reha GmbH, Saarbrücken
Amazon Distribution GmbH, Leipzig
ISBN: 978-3-8381-3106-1

Imprint (only for USA, GB)
Bibliographic information published by the Deutsche Nationalbibliothek: The Deutsche Nationalbibliothek lists this publication in the Deutsche Nationalbibliografie; detailed bibliographic data are available in the Internet at http://dnb.d-nb.de.
Any brand names and product names mentioned in this book are subject to trademark, brand or patent protection and are trademarks or registered trademarks of their respective holders. The use of brand names, product names, common names, trade names, product descriptions etc. even without a particular marking in this works is in no way to be construed to mean that such names may be regarded as unrestricted in respect of trademark and brand protection legislation and could thus be used by anyone.

Cover image: www.ingimage.com

Publisher: Südwestdeutscher Verlag für Hochschulschriften GmbH & Co. KG
Heinrich-Böcking-Str. 6-8, 66121 Saarbrücken, Germany
Phone +49 681 37 20 271-1, Fax +49 681 37 20 271-0
Email: info@svh-verlag.de

Printed in the U.S.A.
Printed in the U.K. by (see last page)
ISBN: 978-3-8381-3106-1

Copyright © 2012 by the author and Südwestdeutscher Verlag für Hochschulschriften GmbH & Co. KG and licensors
All rights reserved. Saarbrücken 2012

List of Papers

This thesis is based on the following papers:

Sauter, T., C. Schneider, R. Kilian and M. Moritz (2009): Simulation and Analysis of Runoff from a partly glaciated meso-scale Catchment area in Patagonia using an Artificial Neural Network. - *Hydrological Processes*, 23 (7), 1019-1030. (c) John Wiley and Sons.

Sauter, T., B. Weitzenkamp and C. Schneider (2009): Spatio-temporal prediction of snow cover in the Black Forest mountain range using remote sensing and a recurrent neural network. - *International Journal of Climatology*, 30 (15), 2230-2341, doi: 10.1002/ joc.2043. (c) John Wiley and Sons.

Sauter, T. and V. Venema: Natural three-dimensional predictor domains for statistical precipitation downscaling. *Journal of Climate*, 24 (23), 6132-6145. doi: 10.1175/2011JCLI 4155.1. (c) American Meteorological Society.

The author of this thesis was responsible for the data preparation, simulations, analysis and writing in all papers.

Contents

List of Papers	i
Abstract	1
Zusammenfassung	3

1 Introduction   5
    1.1 Motivation 5
    1.2 Nonlinearity 6
    1.3 Stationarity and noise 7
    1.4 Modelling issues 7
    1.5 Objectives and aims 8

2 Simulation and Analysis of Runoff from a partly glaciated meso-scale Catchment Area in Patagonia using an Artificial Neural Network   11
    2.1 Introduction 12
    2.2 Methods 13
       2.2.1 ANN 13
       2.2.2 Global Sensitivity Analysis (GSA) 16
    2.3 Study site and data 18
    2.4 Results and Discussion 23
       2.4.1 Evaluation 23
       2.4.2 Sensitivity Analysis 27
    2.5 Conclusion 31

3 Spatio-temporal prediction of snow cover in the Black Forest mountain range using remote sensing and a recurrent neural network   33
    3.1 Introduction 34
    3.2 Study Area 35
    3.3 Data 38
       3.3.1 MODIS satellite data 38
       3.3.2 Meteorological data 38

Contents

- 3.4 Methods 39
  - 3.4.1 Nonlinear AutoRegressive network with eXogenous inputs (NARX) . 40
  - 3.4.2 Fractional snow cover mask 42
  - 3.4.3 Interpolation of snow days 43
- 3.5 Results 43
  - 3.5.1 Fractional snow mask 43
  - 3.5.2 Present situation 44
  - 3.5.3 Predicting future snow cover days 50
- 3.6 Summary and conclusions 54

4 Natural three-dimensional predictor domains for statistical precipitation downscaling   **59**
- 4.1 Introduction 59
- 4.2 Predictor selection 63
  - 4.2.1 Self-organizing maps 63
  - 4.2.2 Simulated Annealing 64
- 4.3 Global Sensitivity Analysis (GSA) 66
- 4.4 Case study: Rhineland region 67
  - 4.4.1 Data and set-up 67
  - 4.4.2 Predictor domains 68
  - 4.4.3 Air masses 72
  - 4.4.4 ANN Downscaling 76
  - 4.4.5 Global Sensitivity Analysis 77
- 4.5 Conclusions and Discussion 80

5 Discussion and conclusions   **85**
- 5.1 General modelling issues 85
- 5.2 Nonlinear determinism 87
- 5.3 Predictor optimization 88
- 5.4 Global sensitivity analysis 88

A Fourier based surrogates   **91**

B Locally constant predictor in phase space   **93**

C Snow-cover maps   **95**

D Estimated monthly changes in the number of snow-cover days   **103**

List of Figures   **129**

List of Tables   **133**

Bibliography 135

Acknowledgements 145

Nomenclature

AC	Anomaly Correlation
ACF	Auto Correlation Function
AIC	Akaike'sches Informationskriterium
ANN	Artificial Neural Network
ASCE	American Society of Civil Engineers
AWS	Automatic Weather Station
BIC	Bayes'sches Informationskriterium
CC	Cross Correlation function
CWB	Climatological Water Balance
DWD	Deutscher Wetterdienst
FAST	Fourier Amplitude Sensitivity Test
GCM	General Circulation Model
GCN	Gran Campo Nevado Ice Cap
GIS	Geographic Information System
GSA	Global Sensitivity Analysis
IAAFT	Iterative Amplitude Adjusted Fourier Transform
IPCC	Intergovernmental Panel on Climate Change
MF	Wet scenario
MLR	Multiple Linear Regression
MODIS	Moderate Resolution Imaging Spectroradiometer
MSE	Mean Squared Error
MT	Dry scenario
NARX	Nonlinear AutoRegressive network with eXogenous inputs
NDSI	Normalized Difference Snow Index
NDVI	Normalized Difference Vegetatio Index
NN	Neural Network
NSIDC	National Snow and Ice Data Center
PACF	Partial Auto Correlation Function
PDF	Probability Density Function
RMSE	Root Mean Squared Error

Abstract

This thesis addresses important aspects in model development and evaluation of nonlinear non-parametric data-driven hydrological and climatological prediction models. Limitations and caveats of data-driven algorithms are discussed using two test cases. A static neural network is developed to forecast the runoff of a meso-scale, partly glaciated, alpine catchment area in the southernmost Andes in Patagonia. With an example of snowcover prediction in the Black Forest mountain range issues of stability and error propagation of dynamical neural networks are discussed. Results are evaluated and compared to simple linear methods. Such algorithms are extremely efficient even if knowledge of underlying processes is missing. Since no phenomenological meaning can be assigned to internal model parameters it is difficult to make causal inferences on the predictors. To overcome this issue we propose to estimate different sources of uncertainty in the model input by a global sensitivity analysis. This approach captures the interaction effects in the predictor set which is in particular an important characteristic of nonlinear systems. Based on this knowledge irrelevant predictors can be pruned, thus effectively reducing the number of predictors for more parsimonious models. Further a novel predictor optimization algorithm for precipitation downscaling which allows for nonlinearities in the screening process is presented. The algorithm optimizes both, the predictors and their corresponding domains by self-organizing maps and a simulated annealing algorithm. Due to the nonlinear screening data-driven algorithms significantly improve the ability to capture complex spatio-temporal structures.

Zusammenfassung

In der vorliegenden Doktorarbeit werden wichtige Aspekte in der Modellentwicklung und Optimierung von datenbasierten hydrologischen und klimatologischen Vorhersagemodellen diskutiert. Vor- und Nachteile von datenbasierten Algorithmen werden anhand zweier Beispiele näher untersucht und denen einfacher linearer Verfahren gegenübergestellt. Zum einen wird unter Anwendung eines statischen Neuronalen Netzwerkes der Abfluss eines vergletscherten mesoskaligen Einzugsgebiets im südlichen Patagonien modelliert und analysiert. Zum anderen werden die Stabilität und Fehlerfortpflanzung dynamischer Netze am Beispiel der Schneedeckenvorhersage im Schwarzwald behandelt. Auch ohne genaue Prozesskenntnis der zu modellierenden Phänomene liefern beide Algorithmen sehr gute Ergebnisse bei der Rekonstruktion von komplexen raumzeitlichen Strukturen. Da jedoch den internen Modellparametern keine phänomenologische Bedeutung zukommt, lassen sich folglich keine kausalen Rückschlüsse in den Prädiktoren ableiten. Um dieses Problem zu umgehen, werden die Unsicherheiten in den Modelleingangsgrößen mit einer globalen Sensitivitätsanalyse abgeschätzt. Dieser globale Ansatz berücksichtigt unter anderem Interaktionen von Prädiktoren, welche vor allem bei nichtlinearen Systemen eine besondere Rolle spielen. Anhand dieser Unsicherheiten lassen sich schließlich kausale Zusammenhänge in den Eingangsgrößen ableiten. Diese Informationen bieten weiterhin einen guten Hinweis zur effektiven Reduzierung der Prädiktoren. Schließlich wird ein neuer nichtlinearer Algorithmus zur Optimierung von Prädiktoren für ein Niederschlags-Downscaling präsentiert. Die Prädiktoren sowie die dazugehörigen Domänen werden mit Selbstorganisierenden Merkmalskarten und einem *Simulated Annealing* optimiert. Der Vergleich mit einem linearen Verfahren zeigt, dass die Vorhersagequalität allein durch eine nichtlineare Selektion von Prädiktoren signifikant verbessert werden kann.

Chapter 1

Introduction

1.1 Motivation

In Geosciences inferences on processes are usually derived from irregularly measured data in both space and time. Unlike laboratory experiments with approximately constant boundary conditions, such empirical time series are characterized by changing boundary conditions, nonlinearity and uncertainty (noise) that might cause misleading results. Hence, phenomenologial knowledge of processes is indispensable in order to derive physically meaningful equations describing the underlying dynamical system. Despite the vast number of analytical tools, the identification of processes is the most time consuming step in model development. In general, unambiguous inferences prove to be difficult when nonlinearity and non-stationarity are present. To become finally accepted, new theories mostly require a robust verification by means of successful predictions for a large number of test cases. Solutions to many of these problems require either the use of nonlinear processors or appropriate linearization schemes. Numerical models falling back on the latter are still providing the best way of verifying theories. They also have a greater capacity to describe systems under unobserved circumstances. However, if only prognostic solutions are neccessary, numerical model development is time consuming and often too expensive. An alternative approach for this purpose are data-driven models. Such models are specifically built to be parsimonious with a minimal set of adjustable parameters, intended to reproduce the statistical properties of the signals. In the domain of atmospheric science the approximation of complex processes by data-driven methods have been well established, e.g. LES simulation (Moreau et al., 2006; Sarghini et al., 2003), downscaling (Haylock et al., 2006; Wilby et al., 1998), probabilistic climate change projections (Knutti et al., 2003), pattern recogni-

1 Introduction

tion (Crane and Hewitson, 2003; Hong et al., 2004) and parametrization schemes (Chevallier et al., 2000; Krasnopolsky et al., 2005). A comprehensive list of references on nonlinear system identification and its application in signal processing, communications, and engineering is given by Giannakis and Serpedin (2001).

The modelling process is influenced by the previously mentioned preconditions and eventually by the demand on application. With respect to modelling issues three, mostly conflicting, measures of model utility can be formulated: (i) approximation accuracy, (ii) physical interpretation and (iii) ease of development. Thus, the choice of the model type is always a trade-off and primarily depends on the practical requirements and general needs.

This dissertation aims to make a contribution to resolve the discrepancy between prediction accuracy, physical interpretation and model development at low cost. Different aspects in the modelling process are worked out and predicted time series are evaluated with regard to nonlinear deterministic structures. Based on uncertainty estimation of static and dynamic neural networks, an attempt is made to detect coherences and structures in the predictor set. In the following section, some important characteristics of nonlinear non-stationary time series and the related modelling issues are discussed.

1.2 Nonlinearity

Since Newton invented differential equations in the mid-1600s (Newton, 1999), it gave scientist an effective way to describe the evolution of dynamical systems in continuous and discrete time (iterated maps). Eventually, the successful application to even complex systems, e.g. the two-body problem, made differential equations a widely used approach in physics, mathematics, meteorology and biology. Corresponding to the number of independent variables, two types of differential equations are distinguished: ordinary- and partial differential equations. While ordinary differential equations involve only time as independent variable (e.g. damped harmonic oscillator), partial differential equations consider both time and space (e.g. heat equation). If such systems include higher order terms such as higher exponents or functions, the system is denoted as nonlinear. At first glance, this does not seem to be a big drawback, but in most cases, the inherent nonlinear interactions make it impossible to solve such problems analytically (Kantz and Schreiber, 2004; Strogatz, 2001; Alligood et al., 2000). Using appropriate approximations some of these problems might be transformed to linear ones and therefore solved analytically. Linear systems underlie the principle of superposition making it possible to decompose complex problems into several independent sub-problems. The additivity guarantees that the net response of a system is then equal to the sum of the individual solutions. Therefore, all irregular behaviour of the system has to be attributed to some random external processes. Unfortunately, the principle of superposition does not apply for nonlinear problems. Complicated dynamics and structures are also created by apparently simple deterministic systems. This then raises the question of, how information can be extracted from empirical data which does not follow the linear paradigma and thereby improve prediction. Modern nonlinear time series analysis attempts to answer these questions on the theory of dynamical systems and tries to establish the link to empirical data. It is based on a geometric approach in phase space which goes back to the outstanding work of Poincaré in the late 1800s (Kantz and Schreiber, 2004; Strogatz, 2001; Alligood et al., 2000). The geometric view provides methods to derive invariant quantities, such as Lyapunov exponents (Grassberger and Procaccia, 1983; Rosenstein et al., 1993) and correlation dimensions (Grassberger and Procaccia, 1983; Schreiber and Schmitz, 2000), characterising the evolution and structure of time series. Improved knowledge about the dynamical nature of measured data is eventually the path to a successful model development.

1.3 Stationarity and noise

The sensitivity of statistical quantities and methods to noise and stationarity often makes their application to real life problems difficult. Stationarity implies that all joint probabilities between states of the system are independent of time within the entire observation period. This general definition also applies to deterministic rules when such rules are interpreted as simple Markov-Chain with transition probability of one (Wilks, 2006; Kantz and Schreiber, 2004). Due to the lack of observations, most climatological time series appear to be non-stationary, although in the limit of infinitely observations they might not be so. In fact, insufficient process understanding prevents the possibility to know for certain that joint probabilities are indeed constant over time. In the same way we cannot make reliable statements on the amount of noise overlaying the signal. Measurements and dynamical noise usually show random perturbations which additionally hamper the decomposition of signal and noise. Whether an algorithm or model successfully predicts the deterministic part of a system is often limited by the amount of noise.

1.4 Modelling issues

Scientists pursue the objective of deriving physically meaningful equations that describe the underlying processes. In case no phenomenological conclusions can be drawn from measurements or if the understanding of the processes is limited, such relationships become difficult to establish. Nonparametric models based entirely on time series data have proven to be very efficient, even for incomplete data that do not provide unique solutions of the inverse problem (Kantz and Schreiber, 2004; Fan and Yao, 2005). Unfortunately, missing phenomenological meaning of the model parameters are the inevitable consequence of this simplification. In the simplest case, linear correlation such as nonzero autocorrelation allows the prediction of future values by linear combinations of the preceding data (e.g. autoregressive moving average model). Nonlinear sources of predictability, however, are characterized by exponentially fast decaying autocorrelation functions so that predictions cannot rely on this information. Nevertheless, nonlinear determinism is very likely to be present in most systems even though irregular structures often mask a clear signal. As long as a deterministic component is inherent, we might approximate future values even in an apparently chaotic (irregular) signal using nonlinear prediction algorithms such as neural networks, radial basis functions or local methods in phase space. Such prognostic data-driven models are usually not affected by the tragedy of equifinality (different models yield the same results), as the phenomenological meaning is missing. Although physical meaning cannot be assigned to model parameters, we do have the possibility of analyzing the complete model structure itself. This is mostly achieved by decomposing the variance of the model output according to the uncertain input factors (Homma and Saltelli, 1996; Saltelli et al., 2000; Sobol, 2001).

1.5 Objectives and aims

This dissertation focuses on the fundamental aspects of uncertainty estimation of nonlinear data-driven prediction methods. Within this framework, the essential factors in the model development process are emphasized and discussed on the basis of both climatological and hydrological time series. A key issue of the thesis is whether the analysis of uncertainties might contribute to process understanding and eventually support the optimization of data-driven models. In particular special attention is paid to predictor optimization with the aim of developing more parsimonious models.

The overall objective is divided into specific aims:

- to implement dynamic and static neural networks to complex nonlinear time series and evaluate their predictive ability (Chapter 2 and Chapter 3),
- to estimate the uncertainty of data-driven models (Chapter 2 and Chapter 4),
- to identify whether data-driven models capture the nonlinear determinism of systems (Chapter 3),
- to develop an efficient predictor optimization for nonlinear prediction algorithms (Chapter 4).

Individual Chapters of this work originate within the framework of independent research projects with fundamentally different objectives. This justifies the different nature of the problems treated which range from glacio-hydrological modelling to precipitation down-

1.5 Objectives and aims

scaling. Consequently, the provided time series broadly differ in their complexity and hence by their inherent properties. In the follwing paragraph a brief description of these time series is given.

Chapter 2 deals with a hydrological time series of a meso-scale alpine catchment area in the southern Andes in Patagonia. The study is embedded in the interdisciplinary Gran Campo Nevado research project. Water levels were recorded by a pressure sensor on a three-hour time resolution over a two-year period. Since the catchment is partly glaciated, energy balance is significantly involved in runoff generation. Processes are widely known thus allowing the prediction of the deterministic component of such systems by first- and second order differential equations. The slowly decaying autocorrelation function indicates that the temporal evolution of the system can also be predicted by linear statistical methods. Due to the short time series, dynamical nonstationarity can be excluded. However, a slow drift by measurement errors as well as noise by flow turbulence superimposing the real runoff signal must be assumed.

Applications to strong nonlinear and nonstationary time series are presented in Chapter 3 and Chapter 4. Both studies are concerned about precipitation time series either in the form of snow (Chapter 3) or liquid water (Chapter 4). The general statistical characteristics of these data are similar and formation processes depend on the same physical variables. Unfortunately, it is impossible to measure these variables on all relevant scales in order to make reliable predictions. Even if high resolution data would be available, prediction is still limited by stochastic components. Since the description of these processes by differential equations turns out to be difficult or even impossible, linear and nonlinear parametrization schemes are usually used. Besides the fast decaying temporal autocorrelations, statistical model approaches have to deal with anisotropic spatial correlations changing with atmospheric conditions.

Chapter 2

Simulation and Analysis of Runoff from a partly glaciated meso-scale Catchment Area in Patagonia using an Artificial Neural Network

In this study a model based on an artificial neural network (ANN) was developed to forecast the runoff of a meso-scale, partly glaciated (40%), alpine catchment area in the southernmost Andes in Patagonia, Chile. The study area is located in a maritime climate with a mean annual air temperature of +5.7°C and about 5500 mm of precipitation per year at sea level. The multilayer feed-forward network is designed to make use of the Levenberg-Marquardt algorithm to increase the speed of computation (convergence). Using climate data recorded at an automatic weather station nearby as well as water level records measured simultaneously, the ANN model was trained and verified using independent training and validation data sets. Parameters and the corresponding time lags were determined by statistical methods such as cross-, auto-, and partial-auto-correlation. The results of the simulation confirm that the proposed model was able to identify the underlying nonlinear relationships between the input parameters and the observed discharge. The correlation during validation shows a significant correlation coefficient of 0.98 and an RMSE of 0.02 m respectively. However, it is almost impossible to decipher the internal behaviour of ANN due to its black-box character. Nevertheless, valuable insights were gained in the complex input-output relationships and the occurrence of dependencies between different input variables were detected using global sensitivity analysis (GSA). The results of the GSA were compared with those of multiple linear regression (MLR). While the performance of the ANN is much better than the MLR, both models return similar results in terms of the dependency of the discharge upon input variables. It was found that despite the large proportion of glaciated surface area within the catchment, discharge is mainly controlled by precipitation (49%). Furthermore, the runoff is slightly influenced by temperature (19%), global radiation (15%) and wind speed (16%). While the ANN proves to be a very efficient tool for simulating runoff in glacerized, alpine catchments from meteorological data, the GSA method as outlined and used in this paper offers a useful approach of analyzing ANN output.

2.1 Introduction

During the last decade informatic tools, such as artificial neural networks (ANNs), have gradually been introduced into geophysical sciences and engineering. Since then ANNs have been successfully applied in hydrological sciences to a wide range of application areas such as classification, sediment transport, forecasting device and submodels of complex processes. A good review on different applications of ANN used for various hydrological problems has been given by Abrahart et al. (2004) and the American Society of Civil Engineers (ASCE) Task Committee. The concept of neuronal computing thereby offers an important alternative to the existing traditional, more well-established, hard computing paradigms (Kecman, 2001; Solomatine et al., 2008). Decisive advantages of the ANN approach mainly arise from its universal ability to approximate any multivariate function, which allows its application to unknown complex systems or processes (Hsu et al., 2005; Tsoukalas et al., 1997; Zell, 2003). Recently, such approaches have become more important in hydrological sciences as research questions often address complex systems that are mostly controlled by nonlinear processes. These nonlinearities induce small uncertainties and unpredictable variations of the system behaviour in time (chaotic behaviour) and cannot be described by classical statistical procedures (Poddig and Sidorovitch, 2001). Adaptation of computation rules through learning even makes networks insensitive to changes in model assumptions and fluctuating boundary conditions so that networks can easily be applied to various problems (Kasabov, 1998; Kecman, 2001). However, despite the multiplicity of so-called data-driven methods and their racy development, a certain reluctance to apply them has been revealed (Abrahart et al., 2004; Maier and Dandy, 2000; Dawson and Wilby, 2001; Jain et al., 2004). This attitude could be ascribed to the general problems of interpreting ANNs, in particular their internal functional behaviour. This implies that it is often not possible to reveal why an ANN produces the answer it does. The knowledge of the network is distributed over the multitude of neuron weights (associative memory) and is hard to reconstruct (Abrahart et al., 2004). The need for explanation triggered several studies dealing with the subject in the last few years: An attempt has been made to derive rules by comparing weight values with parameters of conceptual (Jain et al., 2004) and physical models as well as by heuristic approaches (Sudheer and Jain, 2004; Wilby et al., 2003). This study not only deals with the prediction of runoff by ANN but also tries to point out existing input-output relations by analyzing the complex structure of neuron linkages as an entity by a Global Sensitivity Analysis (GSA), and thus provides an alternative approach to identify model properties. Consciously, assignments of processes to specific hidden neurons have not been conducted since the distributed information is difficult, if not almost impossible, to decipher and therefore it is hard to allocate physical processes to neurons and weights. As the results in this paper show, this method can be useful for the interpretation and comprehension of the model behaviour. Based on this information conclusions of characteristic features of the basin can be drawn. The suggested methodology has been exemplified by the application of the procedure to a glaciated river basin. To the knowledge of the authors, no further studies have yet investigated the ability of ANNs to predict discharge in glaciated river basins.

2.2 Methods

2.2.1 ANN

There are a category of problems that depend on a huge number of subtle factors which cannot be described by linear algorithms (Freeman and Skapura, 1991; Kasabov, 1998; Kecman, 2001). The human brain has the ability to process such information dynamically and learn from it by reorganizing its structure during cerebration. Similar to biological nervous systems, the artificial neural network is a data processing system composed of a large number of simple, highly interconnected elements (nodes, see Figure 2.1) operating in parallel (Tsoukalas et al., 1997). Each node consists of two parts: the propagation function that sums the weighted inputs and, to a certain extent, a nonlinear filter, the so-called activation function (transfer function), through which the activation signal is passed (Equation 2.1). Usually this activation function is a differentiable sigmoid logistic function with values between -1 and 1 (Demuth et al., 2005; Freeman and Skapura, 1991; Kasabov, 1998; Kecman, 2001; Tsoukalas et al., 1997; Zell, 2003). Mathematically the activation function is given by,

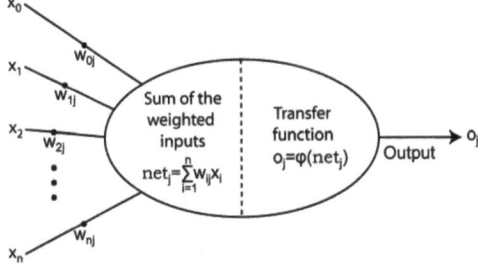

Figure 2.1: Schematic illustration of an artificial neuron.

$$\vartheta(net_j) = \frac{1}{1 + \exp^{-\alpha net_j}} \qquad (2.1)$$

where α defines the slope of the function and net_j the sum of the weighted inputs of the node j (Figure 2.1). The nodes are mostly arranged in three-layer architecture, as shown in Figure 2.2. Information is mainly processed in the hidden layer which at the same time represents the most important unit within the network. The ability of generalization is basically affected by the right choice of nodes in the hidden layer (Dawson and Wilby, 2001; Uvo et al., 2000). Depending on the given input, the output layer finally provides the network result. In addition, some authors (e.g. Kasabov, 1998; Kecman, 2001; Tsoukalas et al., 1997) refer to the inputs, where the data are introduced to the network, as another layer even though they are not real nodes. Besides the network topology, the applied learning algorithms which adapt the connection weights are essential differences between individual networks. This process is accomplished by means of two functions: learning and recall (Tsoukalas et al., 1997). The learning phase adapts the weights with the object of generating the desired output based on the input signal, whereas the recall phase compares the constructed output vector o_j with the desired vector t_j by calculating the network performance function ϵ (Equation 2.2) During training the weights are iteratively adjusted so that the error is minimized. ϵ is usually equal to the Mean Squared Error (MSE) (Demuth et al., 2005), e.g.

$$\epsilon = \frac{1}{n} \sum_{j=1}^{n} (t_j - o_j)^2 \qquad (2.2)$$

The way of adjusting the weights in order to minimize the performance function is given by learning rules. One of the most common rules used for training feed-forward networks is called backpropagation, a supervised learning technique which computes the gradient of the error with respect to the modifiable weights backwards through the network. Instead of using the basic gradient descent methods, the high-performance Levenberg-Marquardt algorithm was applied for fast optimization (minimization) (Demuth et al., 2005). The supervised training process requires an adequate set of learning examples in order to abstract learned data patterns and apply them to unseen examples properly. After successful training (adaptation of weights) the network is able to solve similar problems of the same class that have not been trained explicitly. The quality of simulations depends highly on both the input variables and the architecture of ANNs. Unfortunately, capable procedures to determine optimum network architectures, especially to detect optimal number of nodes in the hidden layer, could not be established so far (Poddig and Sidorovitch, 2001). Basically, the size of the hidden layer just comes up to a fraction of the input layer. Indeed, comprehensive studies of Maier and Dandy (1998, 2000) have shown that the number of neurons can be optimized effectively by simple trial-and-error techniques. Commencing with one neu-

2.2 Methods

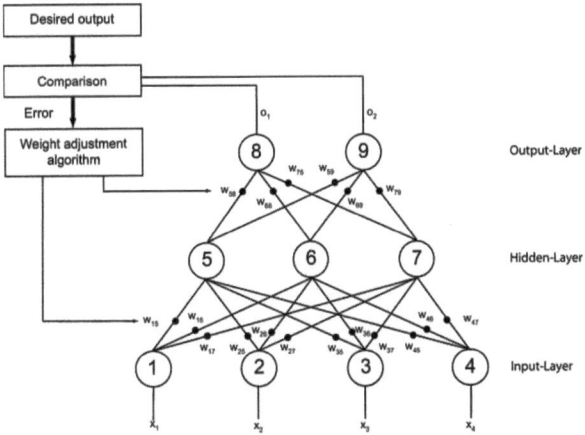

Figure 2.2: Architecture of a multilayer feed-forward network with supervised learning.

ron in the hidden-layer the number of neurons were increased gradually. For each network structure 10 simulation runs were carried out and analyzed in order to get a representative sample. The assessment of the models was undertaken by using the Akaike- (AIC) (Akaike, 1974) and the Bayessche information criterion (BIC) (Rissanen, 1978). These information criteria are computed according to

$$BIC = m \cdot ln(RMSE) + p \cdot ln(m) \qquad (2.3)$$
$$AIC = m \cdot ln(RMSE) + 1 \cdot p \qquad (2.4)$$

where m is the number of samples, p the number of neurones in the hidden layer and the root-mean-squared error (RMSE, see Equation 2.12). The information criteria incorporate statistical performance indicators as well as the complexity of the model, which is indicated by the number of parameters (neurones), in the assessment of the model. The number of parameters is taken into account negatively as otherwise comprehensive models with many parameters are preferred. From Figure 2.3 it follows that, as a result of the great number of input variables, the network requires 47 neurones in the hidden layer. Usually a great number of hidden units (weights) enhances the chance of overfitting. However, the best number of units depends on many factors like the number of input and output units, the number of training cases, the amount of statistical noise, the complexity of the function

2 Simulation and Analysis of river runoff using ANN

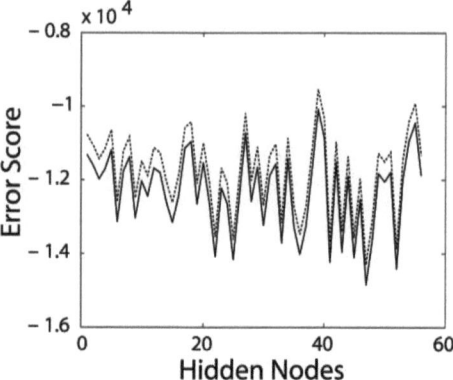

Figure 2.3: AIC and BIC scores for ANN using 1 to 58 hidden nodes.

to be learned and the ANN architecture. There are some rules of thumb which relate the total number of weights of the network to the number of training cases. However, these recommendations do not apply if early stopping (see section on results) is used to avoid overfitting as applied in this study. For the early stopping method it is essential to use many hidden units in order to avoid local optima traps. Therefore, no upper limit on the number of weights is evident until now. The final network architecture consists of 89 neurons in the input layer, 47 neurones in the hidden layer and one neuron in the output layer. All computations were carried out using Matlab software.

2.2.2 Global Sensitivity Analysis (GSA)

Sensitivity analysis provides information on how sensitive a model output is with respect to individual factors of the model, which are usually subject to certain variability or uncertainty (Beven, 2004; Ratto and Saltelli, 2001; Saltelli et al., 2004). Factors can be distinct elements of the model such as parameters or input variables such as in this case study. In contrast to local sensitive analysis, which estimates the partial derivatives of the output caused by a factor whereas nominal values are allocated for the remaining factors, the GSA considers the caused output variability while factors are varied over their domain simultaneously (Saltelli et al., 2000, 2004). For this purpose probability density distributions for each factor have to be defined, from which a sample of factor values is subsequently generated and used for model runs. From these model runs the effect of each factor on the model output can be estimated (Homma and Saltelli, 1996; Ratto and Saltelli, 2001). These effects are best estimated by a variance-based sensitivity analysis since the GSA is neither bound

2.2 Methods

to linear nor non-additive models and, therefore, particularly suitable for studying ANNs (Homma and Saltelli, 1996; Ratto and Saltelli, 2001; Saltelli et al., 2004; Sobol, 2001). Such methods decompose the output variance into factorial terms called importance measures in the form of

$$S_y = \sum_{i=1}^{S} S_{Y|X_i} + \sum_{i} \sum_{i<j} S_{Y|X_{ij}} + S_{Y|X_{1,\cdots,s}} \qquad (2.5)$$

where S is the number of factors, S_Y the total variance of the model output Y and X_i denotes an input factor. The associated importance measures are defined as

$$S_{Y|X_i} = V(E(Y|X_i = x_i^*)) \qquad (2.6)$$

and

$$S_{Y|X_{ij}} = V(E(Y|X_i = x_i^*, X_j = x_j^*)) - V(E(Y|X_i = x_i^*)) - V(E(Y|X_j = x_j^*)) \qquad (2.7)$$

are analogous for higher order terms. The expectation of Y in consequence of X_i having a fixed value x_i^* is described by the term $E(Y|X_i = x_i^*)$ while V stands for the variance over all possible values of X_i. Finally, the variance-based sensitivity indices S_i can be calculated by dividing the importance measures by the total output variance

$$S_i = \frac{S_{Y|X_i}}{S_y} \qquad (2.8)$$

Sensitivity indices with indexes $i = 1, 2, \cdots, s$ are so-called first order indices which describe the main effect of each factor on the model output variance (Saltelli et al., 2004). However, interaction effects and dependencies among factors are not taken into account by the first order indices. Since interaction effects may have a strong influence on the output variance, in particular with an increasing number of factors, the total sensitivity indices S_{T_i} should also be estimated (Ratto and Saltelli, 2001). The total sensitivity indices sum the main effect and all effects of interaction with other factors where a given factor X_i is participating. The S_{T_i} can be computed using,

$$S_{T_i} = \frac{S_{Y|X_i} + S_{Y|X_{i,\sim i}}}{S_Y} \qquad (2.9)$$

where $S_{Y|X_{i,\sim i}}$ indicates all importance measures involving the factor X_i. A method that estimates the main effects as well as interactions between factors is the extended Fourier Amplitude Sensitivity Test (extended FAST) as proposed in Saltelli et al. (2000, 2004) and

Ratto and Saltelli (2001). In addition, this method can treat a group of factors as one single factor, which offers some attractive advantages in application.

2.3 Study site and data

The developed ANN was calibrated with the measured discharge data of the small Rio Lengua in southern Chilean Patagonia on the southern part of Pennsula Muoz Gamero at 52°50'S and 73°00'W (Figure 2.4). Rio Lengua drains the Glaciar Lengua, a minor ice body that is part of the Gran Campo Nevado (GCN) Ice Cap, the southernmost major glaciated area north of the Strait of Magellan. The glacier takes in about 40% of the entire catchment, which measures an area of approximately 15 km^2 (Möller et al., 2007). The basin is laterally bounded by steep flanks which are barely covered with vegetation and predominantly composed of metamorphic bedrock. Hence, the discharge hydrograph is subject to high variability and shows additionally peak discharges. Regional climate can be described as super-humid with annual precipitation amounting to 6500 mm at sea level and even exceeding 10000 mm on higher ground (Schneider et al., 2003). With monthly mean air temperatures ranging from 1.8°C to 9.0°C at sea level, this climate supports heavy glaciations at altitudes above 900 m asl, with fast-flowing outlet glaciers extending down to sea level. These extreme precipitation rates entail high mass-balance gradients (Möller et al., 2007).

In the ablation area of Glaciar Lengua at approximately 450 m asl annual negative mass balance averages to around 8 m/yr and 22 mm/d respectively (Schneider et al., 2007b). Discharge was measured at four gauging stations by means of salt tracer dilution and current meter whereas data from one gauging station have been used (Figure 2.4, Figure 2.5). In addition, meteorological data was gained from an automatic weather station (AWS) located just off the catchment (Figure 2.4) (Schneider et al., 2003). All variables have been recorded with time resolution of 3 hours over a 2-year period, which represents 5048 data points. At first, significant variables were identified on closer examination of the underlying hydrological processes in the catchment causing the runoff. The choice of proper forcing is a critical issue as predictors extensively affect the accuracy of the ANN result.

Due to the large proportion of glacerized ground the energy balance should be considered in the model besides precipitation and antecedent runoff. The energy balance on the glacier is basically affected by the sensible heat flux (54%), net radiation (35%) and latent heat flux (7%) (Schneider et al., 2007b). This behaviour can be particularly explained by high precipitation rates and year-round mild temperatures associated with advection of warm and moist air mass from the Pacific Ocean. Thus, it appears that besides precipitation further predictors such as air temperature, incoming shortwave radiation and wind speed, which account for the discharge from the glacier, should be added as input variables to the model.

2.3 Study site and data

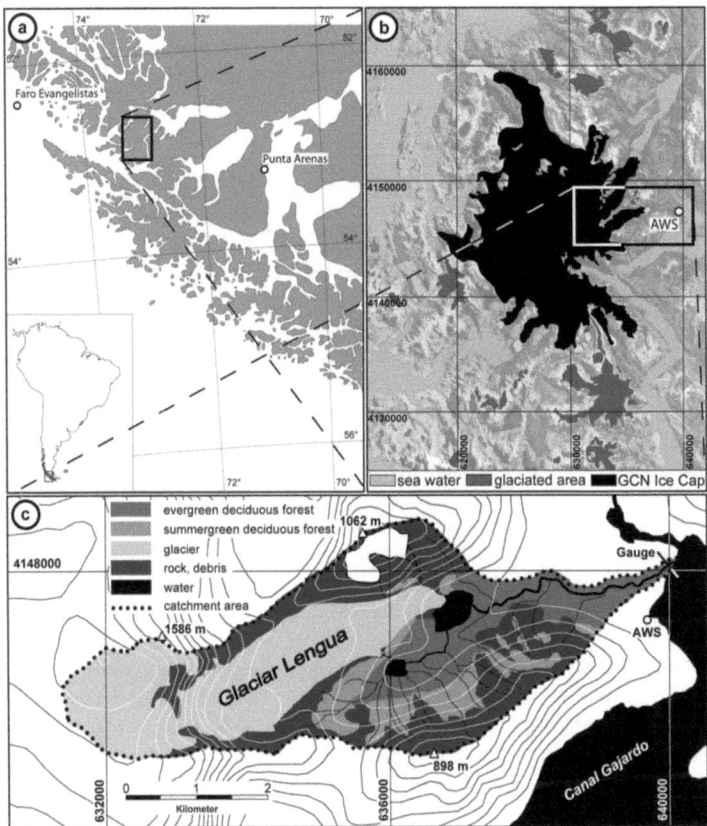

Figure 2.4: Location of the study area and the outline of the drainage basin of Río Lengua in southernmost Patagonia. Coordinates correspond to UTM Zone 18S. The contour lines in (c) are placed at an equidistance of 100 m.

Table 2.1: Description and time lags of the predictors used in the ANN model.

Variable	Description
$runoff_{t-1} \cdots runoff_{t-17}$	Antecedent runoff up to 54 h without actual value
$prec_{t-1} \cdots prec_{t-17}$	Antecedent precipitation up to 54 h
$temp_{t-1} \cdots temp_{t-17}$	Antecedent temperature up to 54 h
$rn_{t-1} \cdots rn_{t-17}$	Antecedent shortwave radiation up to 54 h
$wind_{t-1} \cdots wind_{t-17}$	Antecedent wind speed up to 54 h

Table 2.2: Periods of training, test and validation sets.

Set	Time period	Points	Remarks
Training set	11/04/2002 - 31/05/2003	18-3330	Every second data point (even points)
Test set	11/04/2002 - 31/05/2003	18-3330	Every second data point (odd points)
Validation set	03/06/2003 - 01/01/2004	3376-5048	All data

Adequate time lags of individual variables have been ascertained by qualitative examination of the autocorrelation function (ACF), partial autocorrelation function (PACF) and cross-correlation (CC) curves (Figure 2.6) between influencing variables and the runoff series according to the methodology outlined in Sudheer et al. (2002) and Dawson and Wilby (2001). Basically, this method yields first appropriate estimations, as has been proved by other studies (Jain et al., 2004; Sudheer et al., 2002; Wilby et al., 2003). However, it must be mentioned that classical statistical procedures are not able to trace nonlinear coherences between input and output variables (Poddig and Sidorovitch, 2001). Consequently, the results should be more or less considered as values for orientation. In this context values obtained from the statistical procedure were additionally compared with the experiences of the modeller. The chosen time lags for the used variables as well as the appropriate model input vector are shown in Table 2.1. The different ranges of the variables require rescaling so that their variability, relative to each other, reflects their importance and not their absolute ranges. Therefore, the input and output variables have been scaled so that the values fall within the range [-1,1] (Demuth et al., 2005; Hsu et al., 2005; Kecman, 2001; Shamseldin, 1997; Tsoukalas et al., 1997; Zell, 2003). This advancement leads to better simulation results of extreme peak flows, which are often caused by input values that are out of range (Dawson and Wilby, 2001). To avoid the critical issue of overfitting, the data set was subdivided into separate training, test and validation sets so that the widely used cross-validation technique (Dawson and Wilby, 2001; Abrahart et al., 2004; Solomatine et al., 2008) could be applied (Table 2.2). In this technique the test set is used to monitor the generality of the model during training. Depending on the model performance, the training is stopped if there is no further reduction of the squared error in the test set, while on the other hand the network learns the specific training set. The validation set in contrast consists exclusively of evaluation data and is not used for training at all.

2.3 Study site and data

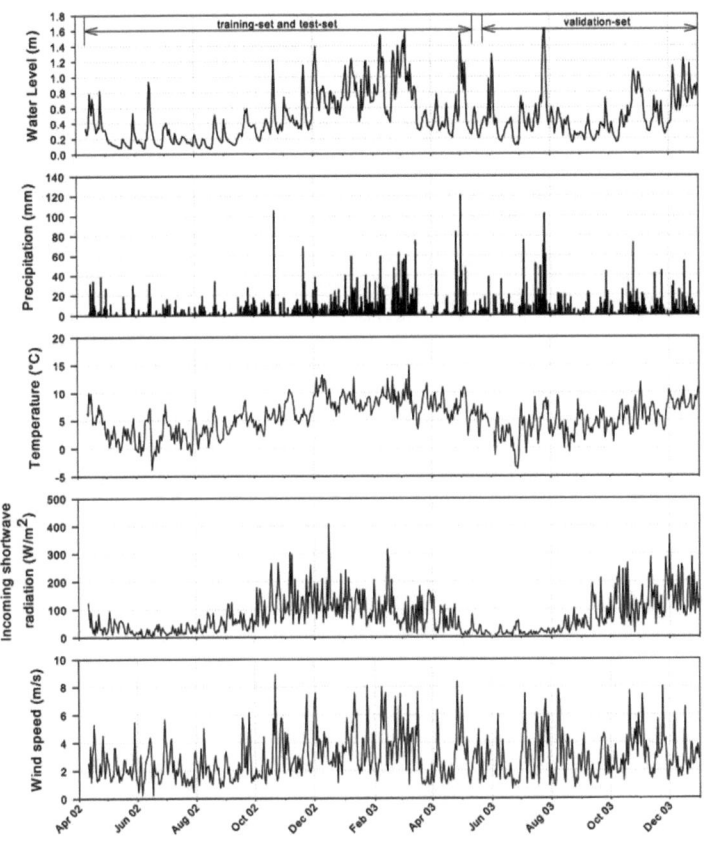

Figure 2.5: Time series of daily mean values and of water level, precipitation, temperature, incoming shortwave radiation and wind speed; daily averages.

2 Simulation and Analysis of river runoff using ANN

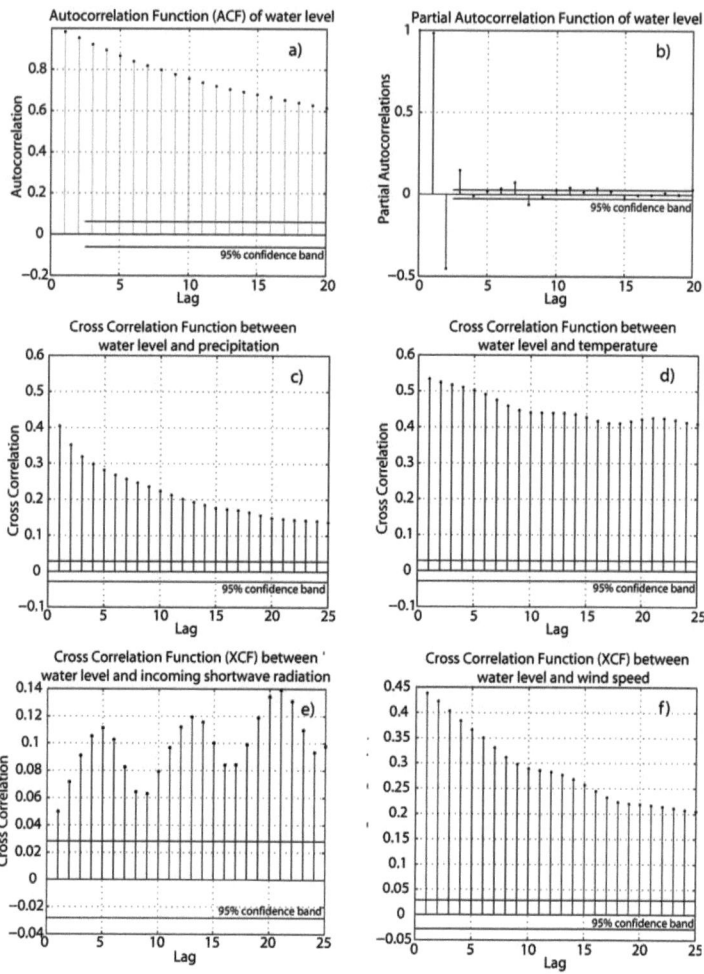

Figure 2.6: Statistical analysis of the input variables: (a) autocorrelation plot of the flow series, (b) partial autocorrelation plot of the flow series, (c) cross correlation plot of the runoff-precipitation series, (d) cross-correlation plot of the runoff-temperature series, (e) cross-correlation plot of teh runoff against incoming shortwave radiation series and (f) cross-correlation plot of the runoff against wind speed series.

Using the available data the three sets were constructed taking into consideration the statistical similarity. In particular, this includes similar ranges and variability. Thereby, it was attempted to incorporate all extreme values in the training set so that the whole domain of the predictors was trained.

2.4 Results and Discussion

2.4.1 Evaluation

In this section, the applicability of using ANN to forecast runoff of the Rio Lengua basin is assessed and compared with a multiple linear regression model (MLR). The performances of both models were assessed by model efficiency (R_{eff}) (Nash and Sutcliffe, 1970), root-mean-squared error ($RMSE$) and coefficient of determination (r^2) as follows

$$R_{eff} = \frac{\sum_{i=0}^{n}(y_i - x_i)^2}{\sum_{i=0}^{n}(x_i - \bar{x})^2} \qquad (2.10)$$

$$r^2 = \frac{Cov(x,y)}{Var(x) \cdot Var(y)} \qquad (2.11)$$

$$RMSE = \sqrt{\frac{\sum_{i=0}^{n}(x_i - y_i)^2}{i}} \qquad (2.12)$$

in which n is the number of data elements, x_i and y_i are the observed and computed flows at time step i respectively, and $\bar{x}$ the mean value of the observed flow. The mean squared errors between observed and modelled runoff in the course of the training are shown in Figure 2.7. While the error of the training subset decreases continuously, the test error increased several times. According to the cross-validation criterion the training stopped after 9 iterations before the defined training goal (validation error < 0.001) could be reached. The results of the ANN forecast for the validation subset are shown in Figure 2.7 and Figure 2.9. In Figure 2.9, residuals between observed and forecasted discharge are plotted against water level. It is observed that deviations are remarkably small (< 0.22 m) over the whole domain of water levels. Although deviations increase with higher water levels, an RMSE value of 0.023 m (validation) indicates the ability of the ANN model to predict high water flows with reasonable accuracy. Systematic errors such as overestimation of peak flows are not identifiable. In order to compare the ANN model outcomes, an MLR model was developed and applied to the same data set. Temperature and precipitation were included in the MLR as explanatory variables. The inclusion of further variables, e.g. wind speed and global radiation, do not lead to any improvement of the MLR results. The best result was obtained

2 Simulation and Analysis of river runoff using ANN

Table 2.3: Model performance indices of the ANN and the MLR models for training and validation.

Model	Training			Validation		
	$RMSE$	R_{eff}	r^2	$RMSE$	R_{eff}	r^2
ANN	0.017	0.997	0.997	0.023	0.992	0.992
MLR	0.205	0.642	0.642	0.211	0.450	0.481

using the present precipitation, precipitation with a 30-hour lag as well as temperature with a 9 hour time lag (see Figure 2.8).

The residual plot of the identified MLR model is presented in Figure 2.10. A comparision of the two models makes it evident that the ANN model is much more powerful than the MLR model (Figure 2.9 and Figure 2.10). A systematic overestimation of low flows as well as an underestimation of high peak flows is clearly apparent in the MLR model (Figure 2.10). The poor efficiency of the MLR is also reflected in the goodness-of-fit values, as given in Table 2.3. Both the systematic error and the poor efficiency indicate that a large amount of variance is unexplained.

Besides residual plots, closer analysis of the hydrographs reveals more detailed information on the model behaviour and quality. Figure 2.11 shows a short period of the observed and the computed hydrographs, generated by both the ANN and the MLR, during validation. The ANN mapped precisely all ranges of the hydrograph. Since the increase in discharge and the gradual decay in flow have been mapped successfully it can be concluded that the non-linear coherences of the system have been captured satisfactorily. This result also confirms the right choice of predictors and their associated time lags (Abrahart et al., 2004). However, the forecast of the MLR model is subject to great fluctuations and is not able to predict future events with adequate accuracy. It is obvious that the association between discharge and predictors cannot be explained sufficiently by a linear method such as the MLR. Attention must be paid to remarkable events such as snow fall and flood events associated with infrequent weather patterns. The former can occur if the air temperature falls below +2°C. Based on a temperature gradient of 0.63 K/100 m (Schneider et al., 2007b), snowfall has to be expected on Glaciar Lengua when positive temperatures up to +4°C are measured at the AWS. Consequently, runoff is reduced because of snow retention and discharge becomes relatively low. In contrast, with the onset of snow ablation a marked peak flow has to be expected. As Figure 2.5 shows, temperature temporarily falls below this threshold. Nevertheless the ANN is capable of providing very good forecast of the river flow also during periods with low temperatures (Figure 2.13). In this case the MLR method shows significant limitations due to its linear character. This leads to a systematical underestimation of the runoff during periods of low temperature in the MLR model. Indeed, the MLR model even predicts negative water levels. The evaluation of flood events, which occur more fre-

2.4 Results and Discussion

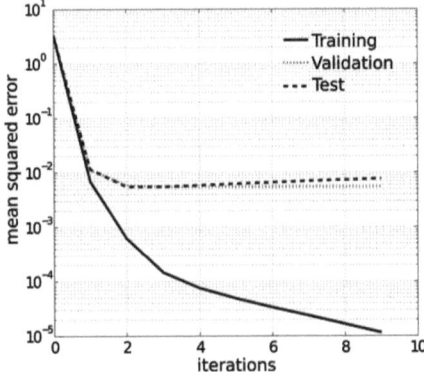

Figure 2.7: Error trajectory of training, test and validation.

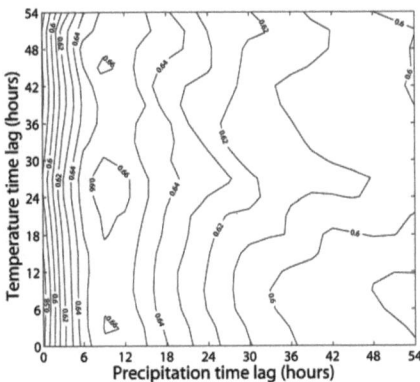

Figure 2.8: Corrrelation coefficient of the MLR model for different time and precipitation time lags.

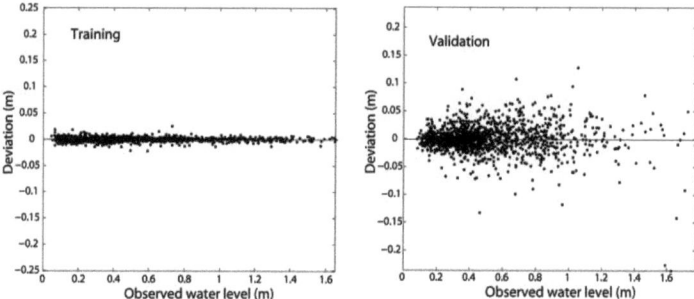

Figure 2.9: Comparison of actual and predicted water levels of the ANN model for training and validation phases.

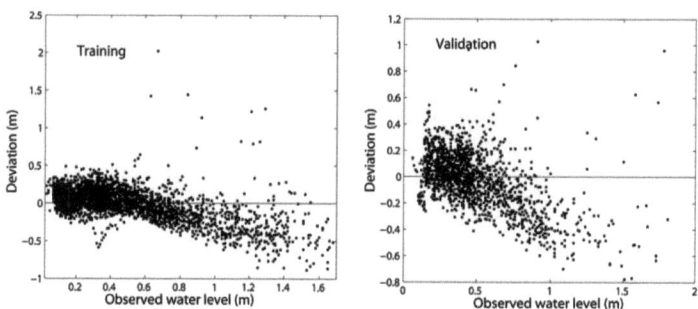

Figure 2.10: Comparison of actual and predicted water levels of the MLR model for training and validation phases. Note that the labeling of the axes differs from Figure 2.9

quently, reveals considerable weaknesses of the MLR (Figure 2.12). Although high water flows are generally underestimated (Figure 2.10), simulated extreme floods are far beyond the measured flow (Figure 2.12). Since specific flood events are mainly affected by precipitation, they are subject to high variability. Generally, this makes great demands on the model since the simulation of such events is crucial for the model quality. Unfortunately, these events are predominantly phenomena that are beyond the range of the training sample and therefore require extrapolation of the model, which is notoriously unreliable for non-linear system behaviours. Nevertheless, the ANN model extrapolated untrained flood events with extraordinary accuracy. The successful mapping of extreme values is probably the result of rescaling the variables into a defined range during pre-processing as described in the previous section. All in all, the ANN model supplies a very good forecast for all parts of the hydrograph whereas the MLR model results are extremely inadequate.

2.4.2 Sensitivity Analysis

The GSA was computed using SIMLAB software. As stated before, density distributions for each input factor must be defined, primarily based on a corresponding variable-set created for simulation. The decision regarding the choice of the probability distribution function (PDF) usually depends on the statistical properties and can possibly be estimated empirically (Saltelli et al., 2004). Since this application is at an exploratory stage, a uniform distribution ranging between their extreme values was assigned to each factor. Estimation of the main and total effects requires approximately 100 ANN model runs (Saltelli et al., 2004) for each input variable so that, in total, 8000 samples were created. Finally, this set of input vectors was applied to the trained ANN model. The training tends to result in different generalization patterns using the same ANN architecture because the weights of ANNs are initialized randomly (between -1 and 1). However, it has been found that these generalization patterns are similar in their structure whereas small fluctuations occur in the rates of each input variable. In order to take into account these fluctuations 10 ANN networks with the architecture mentioned earlier and described in the methods section were analyzed. Consequently, the following sensitivity indices represent the mean values of the sensitivity indices obtained by the analyzed networks. Four indices were combined to form a group for one time, each representing a period of 12 hours. First order effects (see Equation 2.8) explain in total 96.2% of the output variance (S_Y) whereas the remaining 3.8% account for effects of higher order (interaction) (see Equation 2.9).

This ratio demonstrates the strong non-additive character of the ANN and notably simplifies the assignment of effects on the runoff caused by individual variables. It is obvious that antecedent water levels (60 hours) - as represented by the sum of corresponding S_i values - are very present (49.2%) in the computation of the simulated runoff. The percentage of the explained variance in runoff by the antecedent water levels can be associated with the

2 Simulation and Analysis of river runoff using ANN

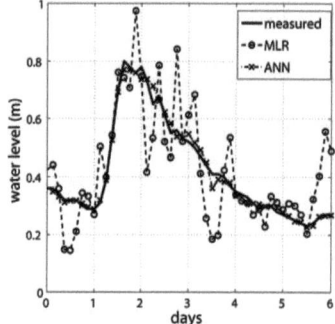

Figure 2.11: Observed and modelled hydrograph of ANN and MLR of a single event (Days 24 Sep 2003 to 30 Sep 2003).

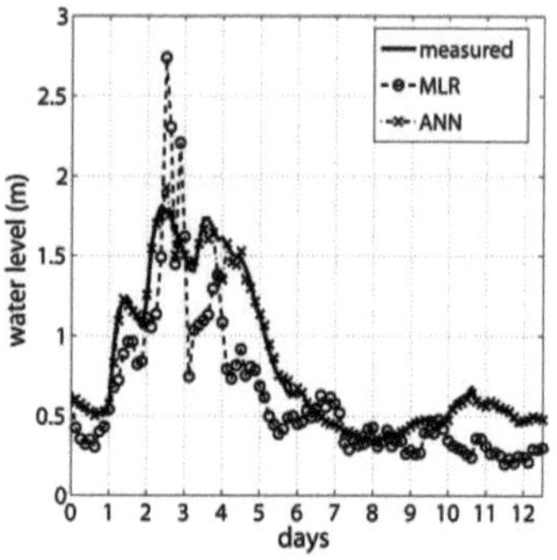

Figure 2.12: Observed and modelled hydrograph of ANN and MLR of an individual flood event (Days 21 Jul 2003 to 3 Aug 2003).

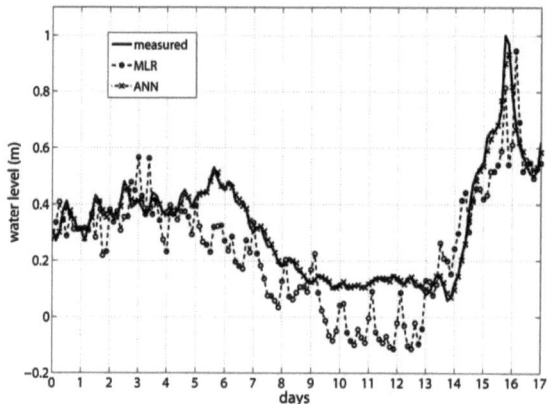

Figure 2.13: Observed and modelled hydrograph of ANN and MLR during a period of air temperature below freezing point (Days 15 Jun 2003 to 2 Jul 2003).

water retained in the catchment. Essentially, this is a function of the catchment properties such as geology, glacier extent, soils and vegetation cover (Abrahart et al., 2004). During periods without precipitation the runoff is just fed by retained water. Hence, antecedent water levels are indispensable for prudential current outflow predictions. Accordingly, the remaining 47.0% of S_Y are attributed directly to the main effects that are caused by the meteorological input variables of precipitation, temperature, incoming shortwave radiation and wind speed. In the further course of this study we concentrate on the effects of the meteorological factors only. Subsequently, percentages presented below refer to this subset of S_Y. Accordingly, Figure 2.14 shows the results of the GSA without taking note of the antecedent water levels. The antecedent 12 hours (four 3-hour mean values) of precipitation explain around 18.7% of the variance caused by meteorological variables. This explained variance decreases when longer antecedent periods of precipitation are considered. This implies that a major part of the precipitation is added directly to the runoff. However, it must be pointed out that the group for 48-60 hours consists of only 2 S_i values while all the other groups consist of 4 values. This extremely decreases the explanatory power of the indices of the last group. Strong interactions between precipitation and other factors can be derived from the high total sensitivity indices of precipitation (Figure 2.15). During the first 12 hours interactions supposedly take place between precipitation and wind speed. The result seems to be reasonable since intensive precipitation in this area is mostly associated with high wind speed (Schneider et al., 2007b). Specific attention must be paid to the influence of temperature on the runoff, as proglacial fluvial systems are usually driven by the diurnal temperature cycle. Despite extensive glaciations such discharge variations superimposed on base flow are not registered. The small effect of temperature is confirmed by minor main effects, which vary between 5.3% and 5.1% up to a time lag of 24 hours. This is in good accordance with the observed hydrograph. Hence, subglacial and supraglacial meltwater draining from the ablation area of the glacier plays a secondary role in the discharge pattern. Furthermore, it must be pointed out that rainfall also contributes to snow and ice ablation due to latent heat input from the atmosphere. This complicates the process of determining the discharge percentage caused by glacial processes. Nevertheless, the importance of this discharge remains far behind the impact of precipitation on the hydrograph. According to the existing precipitation-temperature-distribution the runoff regime can clearly be indicated as pluvio-glacial. Relatively short time lags indicate that channel systems must also be well-developed, despite the fact that the pro-glacial lake diminishes the signal of the temperature. The interpretation of the wind speed and incoming shortwave radiation indices turn out to be more difficult. Despite the highly glaciated drainage basin these variables do not play a decisive role in the runoff formation. Indeed, indices of incoming shortwave radiation are low throughout the entire term although a noticeable periodic variation, which probably displays its daily cycle, is apparent (Figure 2.6 and Figure 2.14). Although wind speed is almost as important as temperature during the first 12 hours (4.8%), the pattern seems rather arbitrary.

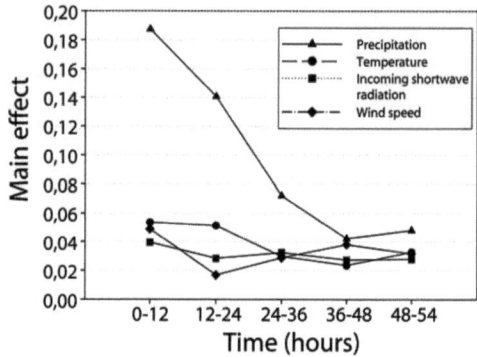

Figure 2.14: Main effects of the meteorological predictors on the runoff (see Equation 2.8).

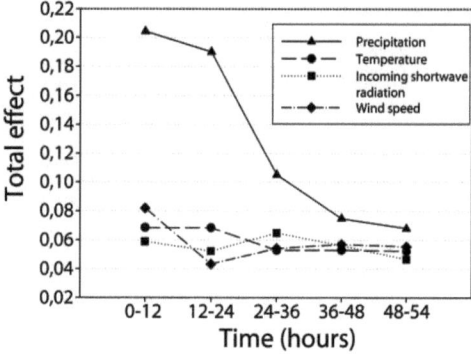

Figure 2.15: Total effects of the meteorological predictors on the runoff (see Equation 2.9).

The knowledge gained from the GSA is qualitatively in good agreement with the prevailing processes as observed in the study area. In contrast, it is difficult to rate the quantitative information obtained as it cannot be evaluated properly. For this reason a comparison with a calibrated physical model could probably provide further detailed information. Since the ANN is able to capture non-linear relations that the MLR is not able to reproduce, it is indicative that rising and falling flows are differently composed by varying impact of the different input variables. Therefore, separate GSAs for each limb of the hydrograph are advisable for further studies.

2.5 Conclusion

This study shows that ANNs are useful tools for analysis and prediction of runoff time series in partly glaciated basins. Furthermore, the ANN model provides by far better results than a simple MLR model. Until recently, no guidelines for developing ANNs could be established and comparisons and assessments of ANNs versus traditional methods are not available in almost all geosciences. A comprehensive comparison with other studies is difficult because of poor information on the data sets used in other studies and different choices of input parameters and network structure. Due to the black-box character of ANNs only the overall impact of driving variables can be directly analyzed within the system. Complex processes cannot be explained by ANNs, which makes them particularly useful in application-oriented studies. More straightforward ways for better understanding of ANNs need to be developed. The GSA proves to be a promising technique in this direction. It is self-evident that the results of the GSA must be judged critically even though the results obtained appear to be realistic. Further investigations on the application of ANNs on other drainage basins are considered to be necessary.

Chapter 3

Spatio-temporal prediction of snow cover in the Black Forest mountain range using remote sensing and a recurrent neural network

Winter tourism is the main economic factor for many different regions in the German Mountain Range. Owing to warming trends experienced in the past and predicted for the future, precise knowledge about the development of snow cover and snow duration in the future is becoming more and more important. On the basis of the Intergovernmental Panel on Climate Change (IPCC) A1B scenario, this paper investigates the possible regional development of snow cover and snow duration in the Black Forest in southwest Germany until 2050. For this purpose, we developed a new method that combines Non-linear AutoRegressive networks with eXogenous inputs (NARX) with Remote Sensing and Geographic Information System (GIS). With this non-parametric approach, we try to define with preferably high accuracy a simple transferable model. Besides the general problem of developing a robust statistical model, our main focus is on the enhancement of the spatial resolution of snow patterns by incorporating complex structures of the underlying terrain using Moderate Resolution Imaging Spectroradiometer (MODIS) satellite data. The results suggest a possible decrease in the number of snow days (snow cover $\geq$ 10 cm) in the decade 2041-2050 by 10 to 44% at altitudes higher than 1200 m, by 17 to 57% at 1000-1200 m and 25 to 66% at 500-1000 m. This results in a dramatic shortening of the snow season mainly caused by earlier snow melt initiation rather than by later rst snow precipitation in autumn. These considerable changes in the snow season would cause enormous losses in the skiing and tourism industries. In this context, the obtained high-resolution snow data and maps can provide a useful tool and decision-making aid for the economy and politics.

3.1 Introduction

Winter tourism is traditionally the most fundamental economic factor in the German Low Mountain Ranges, which signicantly affects the economic prosperity and development of these regions. Wintertime warming trends experienced in recent decades and predicted to increase in the future are mainly driven by an increase in greenhouse gas emissions (IPCC, 2007). As winter tourism is highly sensitive to climate change, this development presents serious challenges for ski areas in these resorts. As a result of the anticipated changes, adaptation strategies have to be seen as a key matter in the near future for the winter sport industries and related businesses. For many such ski resorts, climate change is apparent as a process already underway. In many mountain regions, snow cover was lighter, arrived later in autumn, melted earlier in spring and became more restricted to higher altitudes. These processes and their impact on winter tourism were the subject of many research projects in Central Europe (Breiling and Charamza, 1999; Elsasser and Messerli, 2001; Elsasser and Bürki, 2002) and North America (Hamilton et al., 2003, 2007; Scott et al., 2007, 2008). Recently, many studies of long-term snow cover trends have been published, most of them focusing on regions in Central Europe (Hantel et al., 2000; Beniston et al., 2003; Laternser and Schneebeli, 2003; Falarz, 2004; Hantel and Hirtl-Wielke, 2007). There are only few publications concerning snow cover variability and prediction in the context of climate change and ski tourism for the Black Forest and other German Low Mountain Ranges (Schneider et al., 2005; Schneider and Schönbein, 2006; Schneider et al., 2007a; Schönbein and Schneider, 2005). Although a wide range of snow models exists, many of these studies used classical statistical procedures to estimate changes in snow cover duration and distribution. The persistent popularity of low-order (i.e. second-order) linear models surely arises from the fact that they are easy to develop and that essential model parameters can be obtained analytically. Admittedly, we must keep in mind the inherent nonlinear physical properties and processes of snow, which cannot be described by classical statistical procedures. This basically includes changing thermal and hydraulic properties, changing snow albedo, snow energy balance, heterogeneity of snow cover, formation of snow and snow metamorphism. This leads to the fundamental question: Do linear methods suffciently capture the determinism of the complex system, or can we expect an improvement by considering nonlinearities in statistical models? In recent years, the diversity of nonlinear time series methods encouraged scientists to use such methods for explorative purposes. However, it took many years before such approaches found their way into geosciences and engineering. Currently, predictor design based on nonlinear model structures like articial neural networks (ANN) has been successfully applied in snow prediction ranging from modelling of spatial pattern of snow cover (Tappeiner et al., 2001), retrievals of areal extent of snow cover (Simpson and McIntire, 2001) to real time forecasting of snowfall (Roebber et al., 2007). In most cases, the application of nonlinear methods improved the model quality in terms of model accuracy and model cost. This paper focuses on modelling the snow cover in terms of extent and variability based on atmospheric and topographic conditions for Black Forest Mountain Ranges. The objectives of this paper can be summarized as follows:

- Development of a robust statistical model to predict snow cover in low mountain ranges by combining Nonlinear AutoRegressive networks with eXogenous inputs (NARX) and MODIS satellite data.
- Modelling of snow cover extent at high spatial resolution of 500 m.
- Update of snow cover distribution in the Black Forest Mountain Range until 2050 using future scenarios data from the Potsdam Institute of Climate Impact Research.

In Sections 2 and 3, we introduce the investigation area and meteorological data used for this study. A detailed description of the methods and discriminating statistics are discussed in Section 4. The final model results and model quality are analysed in Section 5. Finally, the paper finishes with conclusions and outlook.

3.2 Study Area

This study focuses on the Black Forest, a German Low Mountain Range located in the southwest of Germany (7.6°-8.8°E, 47.5°-49.0°N) (Figure 3.1 and Table 3.1). Since it is protected from meridional air movement from the south by the Alps, mainly zonal westerly air currents are responsible for local atmospheric conditions (Weischet and Endlicher, 2000). The Black Forest is bordered by the Rhine Graben to the west and south. The highest peak is the Feldberg mountain with an elevation of 1,493 m a.s.l and annual mean precipitation about 1900 mm ((REKLIP), 1995). Precipitation is evenly distributed throughout the year with maximum values of nearly 2200 mm on the windward side of the northern part of the Black Forest, gradually decreasing to values of less than 900 mm on the leeward side. Average annual temperatures vary with altitude between 3.3°C (Feldberg) and 10.1°C (Freiburg i.Br.) ((REKLIP), 1995). The Black Forest is mainly covered with spruce and fir trees. Snow depth has been measured at twelve stations distributed over the whole area of the Black Forest mountain range with a higher concentration of stations in the southern part (Figure 3.1). The lowest station (Pforzheim) is situated at 246 m a.s.l. at the northern end while the highest station is near the summit of the Feldberg at 1,486 m a.s.l. in the southern region. Six stations are situated at elevations of nearly 800 ma.s.l. and above and cover therefore the key areas of the Black Forest with regard to skiing tourism. Four stations represent the mid-altitude mountain region between 500 and 800 m a.s.l. while just two stations are situated in the lowlands beneath 300 m a.s.l.

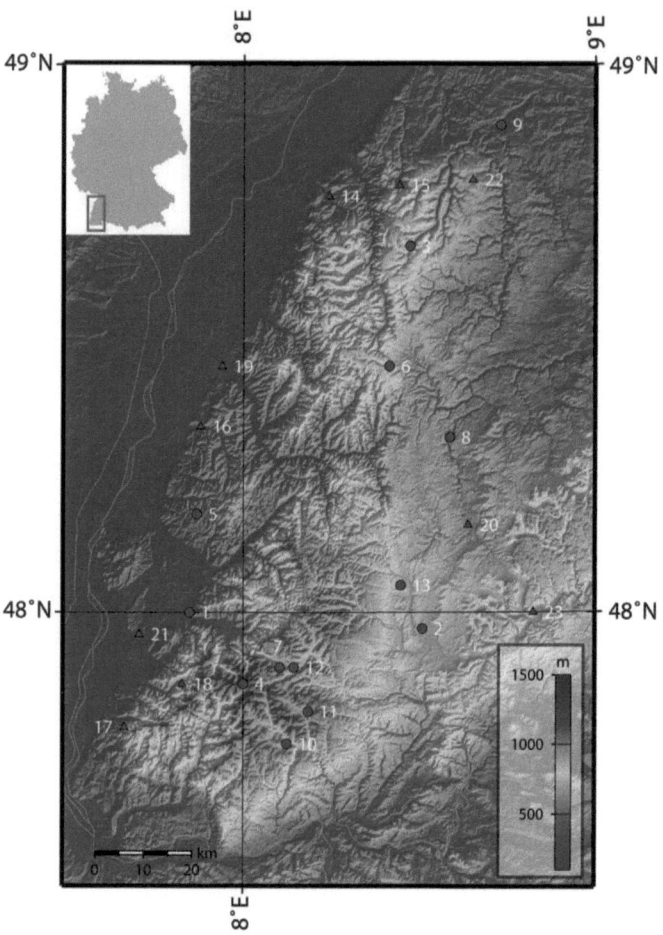

Figure 3.1: Digital Elevation Model of the area of investigation with marked location of each weather station. Circles indicate stations that have been used for prediction whereas the triangles denote additional stations used for interpolation. The inset shows the location of the study area within Germany.

Table 3.1: Location and altitude of the Stations in the study: ID (Figure 3.1), name, geographical coordinates and altitude.

ID	Name	Latitude	Longitude	Altitude (m)
1	Freiburg	48.00	7.85	252
2	Donaueschingen	46.97	8.50	689
3	Enzkloesterle	48.67	8.47	606
4	Feldberg	47.87	8.00	1486
5	Freiamt	48.18	7.87	442
6	Freudenstadt	48.45	8.41	797
7	Hinterzarten	47.90	8.10	886
8	Oberndorf	48.32	8.58	464
9	Pforzheim	48.89	8.73	256
10	Sankt Blasien	47.76	8.12	785
11	Schluchsee	47.82	8.18	960
12	Titisee-Neustadt	47.90	8.14	875
13	Villingen-Schwenningen	48.05	8.44	710
14	Baden-Baden	48.76	8.24	211
15	Herrenalb	48.78	8.44	431
16	Lahr	48.34	7.88	155
17	Muellheim	47.79	7.67	449
18	Muenstertal	47.87	7.83	545
19	Offenburg	48.45	7.94	155
20	Rottweil	48.16	8.63	588
21	Schallstadt	47.96	7.71	213
22	Schoemberg	48.79	8.65	620
23	Tuttlingen	48.00	8.81	648

3 Spatio-temporal prediction of snow cover in the Black Forest mountain range

3.3 Data

3.3.1 MODIS satellite data

MODIS data (Terra product, MOD10A1) with information about snow coverage are offered by the National Snow and Ice Data Center (NSIDC) with daily resolution since 2000. MODIS is a 36-band spectroradiometer covering visible infrared, near-infrared, shortwave-infrared and infrared bands from 0.4-14 μm with a spatial resolution ranging form 250 m to 1 km (Hall et al., 2002). The main objectives are to map global vegetation and land cover, global land-surface change, vegetation properties, surface albedo, surface temperature, snow as well as ice cover at daily resolution. The characteristics of snow are easily detectable with remote sensing tools due to high reflectance characteristics in the visible and high absorption characteristics in the sortwave IR band of the electromagnetic spectrum. To distinguish snow from many other surface features the Normalized Difference Snow Index (NDSI) was established by Hall et al. (1995). The NDSI is analogous to the Normalized Difference Vegetation Index (NDVI). Satellite reflectance values in MODIS bands 4 (0.545 - 0.565 μm) and 6 (1.628 1.652 μm) are used to calculate the NDSI (Hall et al., 2002):

$$NDSI = \frac{band_4 - band_6}{band_4 + band_6} \quad (3.1)$$

To classify a pixel as snow in a moderately forested region the NDSI must be $\geq$0.4 and reflectance in MODIS band 2 (0.841-0.876 μm) > 11%. If the MODIS band 4 reflectance is > 10% the pixel will not be mapped as snow even if the other criteria fit. Changes that occur in the spectra of a forest when it becomes snow-covered are taken into account by additionally using the NDVI, as snow will tend to lower the NDVI. Taken together, NDVI and NDSI provide a strong signal that can be used to classify snow-covered forests (Hall et al., 2002; Klein et al., 1998) like the Black Forest. Based on the NDSI a subpixel snow fraction algorithm using a regression equation was implemented by Salomonson and Appel (2004). The fractional snow cover calculation is applied to the full range of NDSI values (0.0-1.0). All data used in this investigation passed the MODIS Quality Assessment at NSIDC.

3.3.2 Meteorological data

Data from the Potsdam Institute of Climate Impact Research were used for spatially high resolved simulations of future climate developments. Orlowsky et al. (2007) took a resampling scheme for regional climate simulations (STAR II). Similar to the analogue methods, time series are constructed by resampling from segments of daily observations series (1951-2003). This guarantees physically and spatially consistent fields also for variables with high variability like precipitation as the simulated fields are obtained from a pool of observed fields. Instead of using daily data of Global Circulation Models (GCM) as external forcing the resampling method is simply constrained by the simulated annual mean of a characteristic climate variable (usually air temperature) at this location. A large set of simulation ensembles is generated under the assumption of the given linear development of the characteristic variable according to this approach. For each realization the climatological water balance (CWB) is calculated characterized by mean and trend at the representative stations. After calculating the CWB the simulations are arranged from wet to dry. Realizations exceeding the 95%-quantile are considered wet scenarios (MF) whereas realizations that fall below the 5%-quantile are considered dry scenarios (MT). All future climatic changes are based on the A1B emission scenario with an average linear air temperature trend of about 0.26-0.28 K per decade (Table 3.2). This scenario is considered a moderate climate warming scenario between the more extreme scenarios B1 (little warming) and A2 (strong warming) (IPCC, 2007). The Regional Climate Model STARII is forced by the ECHAM-5 General Circulation Model. Snow depth data for the area of investigation were obtained from the German Weather Service (Deutscher Wetterdienst, DWD).

3.4 Methods

Interpolation of in-situ measurements are often affected by insufficient information about underlying spatial structures of the considered variables and processes. However, in the case of snow cover extent and variability remote sensing images are a useful tool to identify complex spatial structures as snow is easily detectable in various radiant energy bands. Thus, snow interpolation algorithms can be modified and improved based on this information. Under the fundamental assumption that spatial structures do not vary profoundly in the future and snow depths can be predicted with sufficient accuracy for various sites, it is possible to create snow cover maps for future scenarios on a regional scale with high spatial resolution.

Table 3.2: Yearly means of the predictors for the period 1990-2000 and the simulations of future climate developments at station Feldberg.

	1990-2000	wet scenario		dry scenario	
		2021-30	2041-50	2021-30	2041-50
Temperature (C)	4.03	4.65	5.65	4.76	5.77
Precipitation (mm/yr)	1661	1891	1997	1708	1641
Wind speed (m/s)	8.25	8.12	8.32	8.30	8.08
relative humidity	81.8	81.4	82.7	81.4	79.7
solar radiation (W/m)	1006	1032	1059	1066	1092

3.4.1 Nonlinear AutoRegressive network with eXogenous inputs (NARX)

The method described is based on this approach, which can basically be subdivided into two successive steps. First, snow depths are modelled at different weather station sites by Neural Networks (NN). Second, the modelled snow depths are interpolated by means of fractional snow masks that are derived from MODIS snow maps. The implementation of such a procedure would be straightforward if the system consisted of linear and deterministic processes as these systems allow a rather precise prediction of future values. Unfortunately, the physical and chemical processes within the snow-cover are highly nonlinear and vary both spatially and temporally. This usually leads to a strong dependence on initial conditions whereas small fluctuations in the independent variables can cause great variation in the response of the system. As the system is additionally forced by external processes, which are obviously not constant in the sense of variance and mean during the measurement period, the time series must be non-stationary. It must also be assumed that the data set contains noise, mainly due to measurement errors which can mask the signal to a notable ratio. Such complex time series require a stable prediction model which can deal with nonlinear coherences besides out of sample data (forced by the strong non-stationarity). In this context we claim that the model must reproduce long term dynamics in a statistical sense rather than focus on short term predictions. To ensure a consistent prediction the model must yield an artificial snow cover time series with the same statistical properties as the original data. On the other hand the model should be easily transferable and applicable to other areas of interest at very low cost. In order to fulfil all these requirements a Nonlinear AutoRegressive network with eXogenous inputs (NARX) was chosen to predict snow depths at discrete sites. NARX belongs to the class of recurrent dynamic networks which are commonly used in discrete time-series modelling (Connor et al., 1994). Unlike common backpropagation NN, which are strictly feedforward, NARX have some memory effect forced by propagating data from earlier processing stages to current calculation. Mathematically it can be described by

$$y(t) = f(y(t-1), y(t-2) \cdots y(t-n), \varphi(t-1), \varphi(t-2) \cdots \varphi(t-n)) \qquad (3.2)$$

where $y(t)$ is the model output regressed on previous values $y(t-n)$ with a delay of n timesteps as well as further exogenous parameters φ. The function f is an unknown smooth nonlinear mapping function approximated by the network (Siegelmann et al., 1997; Connor et al., 1994; Lin et al., 1996).

In general, NN are able to approximate arbitrary complex linear and nonlinear dynamical systems without making any assumptions about the underlying data and processes. This brings considerable benefits and simplifications in modelling. Modellers do not have to be concerned about the distribution of data, undetected processes, out of sample input variables, correlations, noise or nonlinearity. On the other hand, the non-parametric char-

3.4 Methods

acter of the model does not provide insights and understanding of the system's behaviour itself and can be treated exclusively as a black box. In this case we hardly get any information about the underlying determinism or even understanding of processes. In recent years some promising attempts have been undertaken in order to derive physical meanings of the model structure, but the analysis of NN is still a challenging task (Sauter et al., 2009). However, since we are interested in the pattern of snow deposition rather than in physical processes this model property is not a major drawback.

Basically, the NARX is used as a predictor of snow depth for the following day. The modelled outcome is fed back to the input layer of the network and will subsequently be used for the next calculations. Depending on the selected training parameters (learning rate, training epochs, training algorithm, number of neurons in the hidden layer, delays) and data set properties this could lead to strong divergence of the generated solution from the true values. Especially, recurrent networks are highly sensible to small changes in the neuronal weights due to the time delays and therefore often tend to produce instabilities, oscillations, bifurcations and chaos (Cao and Wang, 2004). However, a reduction of divergence (error propagation) can be attained by using a so-called serial-parallel architecture during training in which the true output is used instead of the estimated output. As such a network architecture is similar to feedforward networks a static backpropagation algorithm can be used for the adaptation of weights. In case of real simulation the architecture is changed to pure parallel so that the estimated outcomes from the network are used as inputs. Although the risk of overfitting for NARX is rather marginal it is necessary to verify model predictions to an independent validation data set as individual errors can be small but systematic. In this approach training parameters are considered as reliable if the generated time-series is bound to realistic constraints and prediction errors do not increase fast, or even exponentially (chaotic), with time evolution. These criteria must also be warranted if the model is integrated from any arbitrary point of the observed data.

The selection of suitable input variables and their corresponding delays is another crucial factor in the model development. We decided to consider meteorological variables only while time-invariant parameters (i.e. topography) have been disregarded completely because these features are already captured in the fractional snow cover mask. Finally, on the basis of physical process comprehension and trial and error, a set of five variables (air temperature, precipitation, wind speed, humidity, downward solar radiation) are used as input to the network. However, the assessment of the depth of antecedent information is more difficult and strongly depends on the method used. Common linear time series analysis tools do not detect any nonlinear coherences at all which makes them useless in this case. Therefore, we minimized the RMSE by changing the time lag from 1 to 5 days. According to Figure 3.2 all time-lags were set to a fixed value of three days, which should account for sufficient antecedent information. The best network optimization is obtained by using the Levenberg-Marquard algorithm. We used the same NARX architecture at all sites whereas

3 Spatio-temporal prediction of snow cover in the Black Forest mountain range

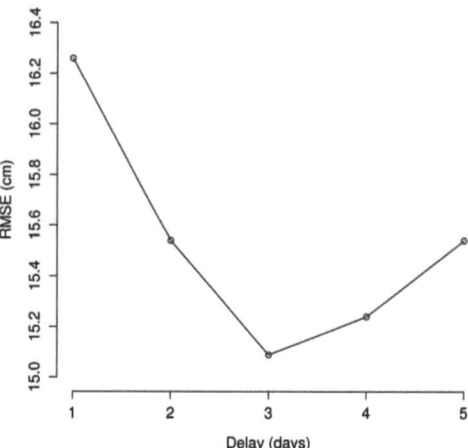

Figure 3.2: RMSE between model output and measured snow depth plotted against the delay of the input variables.

the number of hidden neurons and training epochs have been adapted accordingly by trial and error in order to get the best fit. The quality of the simulations is quantified by the most commonly used error measures such as the mean squared error (MSE) and the root mean squared error (RMSE, see Equation 2.12). Besides the current prediction errors measures we take into account the mean, variance and probability distribution of the time series as these statistics do not involve any correlation in time but rather capture the overall characteristic of the time series.

3.4.2 Fractional snow cover mask

At this processing stage measured and modelled snow depth time-series are available only at discrete sites where in-situ measurements have been accomplished. In the next step the generated snow depths are spatially interpolated on the basis of the information given by the fractional snow cover mask. This mask represents the average fractional snow cover during the winter skiing season, lasting from the beginning of November until the end of April. Statistical performance is done by implementing a cube processing algorithm. In a first step, daily MODIS snow cover masks (Terra product, MOD10A1) reaching from 2001 to 2007 are aggregated to a three-dimensional time-space cube. With this cube, calculations of monthly average values and seasonal average values are realized on a pixel-by-pixel basis. Each pixel finally contains a single value representing the average over the whole investigation period and can therefore be treated as a two-dimensional object. The created image shows the average fractional snow cover over the whole winter season for each pixel (Figure 3.3).

3.4.3 Interpolation of snow days

The number of snow days (days with snow depth > 10 cm) at each station have to be linked to the mean fractional snow cover which was obtained before from the MODIS snow cover product. The interpolation has been realized by fitting a power transfer function

$$f(x) = ax^n \qquad a, n \in \mathbb{R} \qquad (3.3)$$

to the given set of points by minimizing the sum of squares of the residuals. This proceeding allows a spatial interpolation of the snow days by means of the fractional snow mask. In order to reduce variability we rather used the number of snow days than the snow depth for this fit (Figure 3.6). Eventually, for each pixel the number of days can be allocated by means of the derived transfer function. Snow days were defined as days with a minimum snow depth of 10 cm. Once these snow days are filtered out the average number of monthly mean snow days per pixel is calculated.

3.5 Results

3.5.1 Fractional snow mask

The subpixel fractional snow masks clearly display the topography of the Black Forest with the highest snow coverage values of 37% around the region of the Feldberg summit (1.493 m a.s.l.) and the lowest values, ranging from 0 to 3% in the Rhine Valley (~200-250 m a.s.l.) (Figure 3.3). On average, the southern part of the Black Forest is characterized by a higher variability together with higher mean values, which is mainly based on higher mean elevation and a rougher surface structure whereas the northern part of the Black Forest is smoother and lower in altitude with relatively small areas reaching elevations of more than 750 m a.s.l. and just a few summits exceeding 1000 m a.s.l. (Hornisgrinde: 1164 m a.s.l.). Moreover, topographic depressions like river valleys are well reflected in the snow mask. Due to varying cloud coverage a pixel-based sample size for cloud coverage was calculated and integrated into further analysis with the highest sample sizes (up to 529 samples) at higher topographic regions and the lowest sample sizes (up to 392 samples) at lower topographic regions caused by weather situations with stratus clouds at low altitudes.

3 Spatio-temporal prediction of snow cover in the Black Forest mountain range

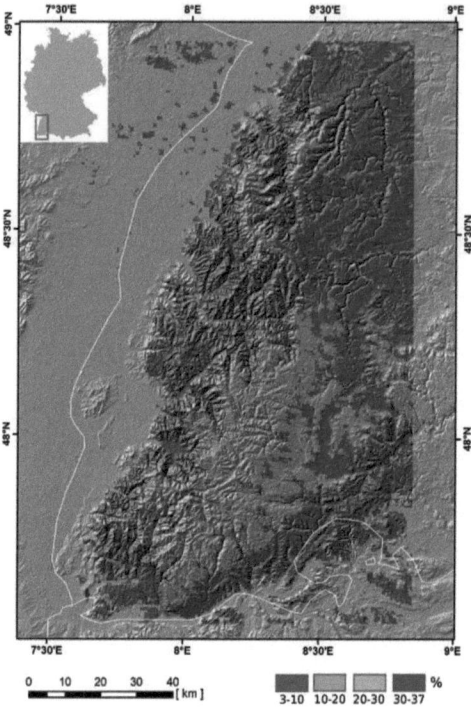

Figure 3.3: Mean fractional snow mask (%) for skiing season (October-April) derived from the MODIS satellite data. The insert shows the location of the study area within Germany.

3.5.2 Present situation

In order to supervise the development of the model training as well as the generalization process the data sets have been divided into suitable validation and training sets. The latter covered January 1951 to December 1993, while the period from January 1994 to December 2003 was used for validation. The validation set was also used as a reference period for changes in future scenarios.

3.5 Results

Table 3.3: Yearly means of the predictors for the period 1990-2000 and the simulations of future climate developments at station Feldberg.

	Measured				Modelled				Error (model)		
	mean	var.	std.	days	mean	var.	std.	days	rmse	logRmse	corr
Donaueschingen	3.40	42.48	6.51	13.7	4.29	53.43	7.30	12.2	5.91	1.77	0.87
Enzkloesterle	8.17	157.40	12.54	28.2	6.23	115.58	10.75	18.7	6.29	1.83	0.91
Feldberg	37.94	858.95	29.30	84.8	33.93	920.33	30.33	74.2	15.09	2.71	0.87
Freiamt	2.42	26.17	5.11	7.5	2.28	21.30	4.61	8.2	2.82	1.03	0.86
Freudenstadt	18.14	557.48	23.61	44.7	16.38	504.20	22.45	40.0	12.15	2.49	0.91
Hinterzarten	19.99	491.53	22.17	53.5	14.81	268.16	16.37	45.1	13.35	2.59	0.88
Oberndorf	3.07	43.81	6.61	11.7	2.36	27.88	5.28	7.0	2.58	0.94	0.93
Pforzheim	0.86	6.52	2.55	4.0	1.06	9.70	3.11	3.2	2.34	0.85	0.83
Sankt Blasien	13.98	336.36	18.34	42.3	9.22	109.81	10.47	27.7	15.14	2.71	0.89
Schluchsee	18.65	526.45	22.94	49.5	16.14	421.92	20.54	39.7	15.38	2.73	0.86
Titisee	16.29	439.77	20.97	45.7	12.01	269.79	16.42	31.8	11.02	2.40	0.91
Villingen	6.45	136.27	11.67	28.6	5.54	75.64	8.69	20.3	7.04	1.95	0.84

45

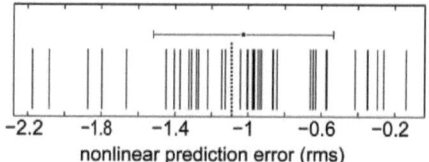

Figure 3.4: Nonlinear prediction error measured for the residuals and the 39 surrogates. The error of the original time series is plotted with a dashed line. The mean and standard deviation of the surrogates is displayed by the error bar.

The convergence criteria and the number of hidden units of the NARX were chosen by cross validation by observing the development of the error measurements of both the training and validation sets (Sauter et al., 2009). From station to station the best results are obtained using approximately 10 to 15 neurons in the hidden unit. This smallness of the number is not due to the complexity but rather to the generalization and stability of the model. Table 3.3 shows the statistical measurements and network performance for the validation period between measured and modelled snow depths. In general the mean snow depths are predicted sufficiently accurately with an average deviation from the mean of about 13% (DJF). Considering the number of snow days the mean error increases by up to 22% (DJF). However, there is a tendency that the NARX has more problems when capturing high snow depths. This is also confirmed by comparing the RMSE and logarithmic RMSE measures (see Table 3.3). On closer inspection the network consistently underestimates the snow depths at all sites. Such a systematic error raises some fundamental questions about the nature of these deviations. In order to prove that the network captures the entire determinism of the system we tested the residuals against the null hypothesis of a Gaussian linear stochastic process. We compared the residuals against surrogate data that shares the same power spectrum and amplitude distribution as the original data (residuals). The surrogate data was generated by an Iterative Amplitude Adjusted Fourier Transform (IAAFT) which was described by Schreiber and Schmitz (1996, 2000) (see Appendix A). In this context the surrogates represent a linear dynamical process which is used to test whether the residuals originate from a nonlinear system (Theiler et al., 1992; Kantz and Schreiber, 2004; Venema et al., 2006a,b). For the test at the 95% level of significance ($\alpha=0.05$) we generated 39 surrogate time series ($2/\alpha - 1$) with the same statistical properties as the residuals mentioned before. We used a locally constant predictor in the phase space (see Appendix B) as a discriminating statistic. The embedding space was constructed using three dimensional delay coordinates at a delay time of one day and a neighbourhood environment of 0.25 times the rms amplitude of the data. As shown in Figure 3.4 the prediction error of the original data is not significantly smaller than for the 39 surrogate time series. This result either suggests that the null hypothesis cannot be rejected and the residuals are a linear stochastic process or that the method is not able to detect the inherent nonlinear structures. As far as this work is concerned, we can assume that the network captures the system structures and that errors are non-systematic on a 95% level of significance. Although the discriminating statistics show a confident result we cannot be sure at the same time that spatial structures are also obtained by the model. For this reason we additionally analysed the cross correlation matrices between the time series of the stations (Figure 3.5). Coherences decrease with increasing distance and weak correlation can be observed between stations on the east and west part of the mountain range (e.g. station 12). Although spatial structures are retained by the artificial time series, correlations between stations are smaller throughout. To affirm these findings we additionally computed the anomaly correlation (AC) which characterizes the spatial accuracy of the predicted time series (Storch and Zwiers, 1999; Wilks, 2006). Mathematically it can be described as

$$AC = \frac{\sum\limits_{m=1}^{M} (y_m - \bar{y})(o_m - \bar{o})}{\sqrt{\sum\limits_{m=1}^{M} (y_m - \bar{y})^2 \sum\limits_{m=1}^{M} (o_m - \bar{o})^2}} \qquad (3.4)$$

where y_m is the predicted mean number of snow days at each station M and o_m accordingly the observed number of snow days. Each station is converted into anomalies by subtracting the long-term mean number of snow days over all stations and $\bar{o}$. The correlation coefficient is bounded by ± 1. The calculated AC value of 0.98 for the winter seasons of the validation set confirms that spatial structures are accurately reproduced by our approach. However, this skill score is not sensitive to biases between the forecasted and modelled fields so that the observed underestimation of the model is not considered. All statistical tests and measurements tend to indicate that the NARX is able to reproduce the observed time series both temporally and spatially. We would expect a further improvement concerning the spatial structure by including the cross correlation matrix or cross-spectrum into the network structure. Unfortunately, within the scope of this work it was not possible to analyse this model structure.

Figure 3.6 exemplifies the fitting of the transfer function (dashed line) to the measured station values (circles) for the months of January and February respectively. All correlation coefficients are significant on a 95% significance level (Student t-test with $\alpha = 0.05$). The triangles display additional stations which could be used for fitting the transfer function even though they were not used for modelling due to a lack of data in the training period. This additional information modifies the transfer function within the lower band of the mean fractional snowcover (solid line). Without this correction the number of snow days at lower and middle elevations would be slightly overestimated. In order to correct each pixel value we constructed an error mask based on the differences observed between the two transfer functions.

3 Spatio-temporal prediction of snow cover in the Black Forest mountain range

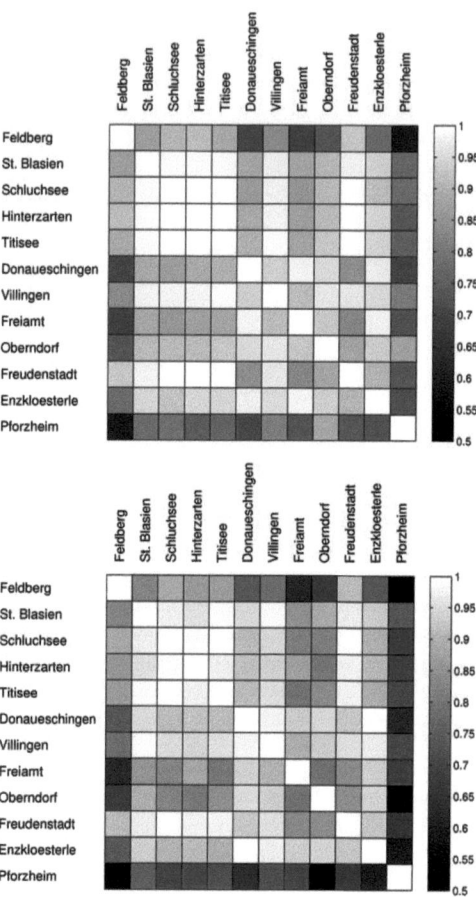

Figure 3.5: Correlation matrices between the original (top) and the simulated (bottom) weather station time series. Locations of weather stations are shown in Figure 3.1.

3.5 Results

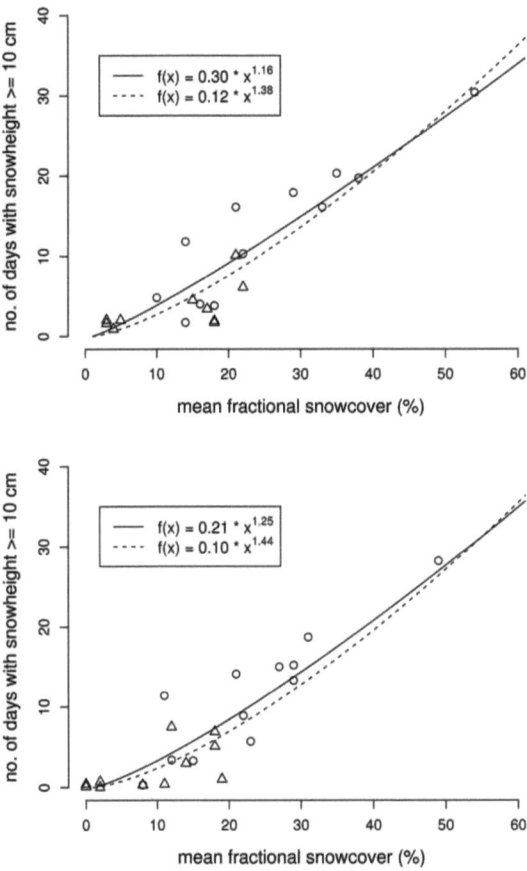

Figure 3.6: Transfer function between the number of days with snow depth greater than 10 cm and the mean fractional snow cover (%) for January (top) and February (bottom). Blank circles indicate stations that have been used for prediction, whereas the triangles are additional stations used for interpolation. Both fittings, with (solid) and without the additional stations (dashed), are shown.

3 Spatio-temporal prediction of snow cover in the Black Forest mountain range

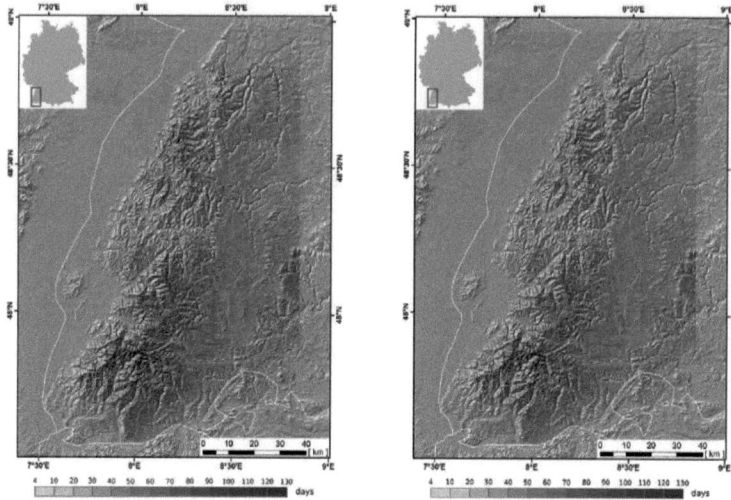

Figure 3.7: The observed (left) and modelled (right) seasonal maps of snow cover days for the validation periods (1994-2003) are displayed. Only heights above 250 m are considered. The inset shows the location of the study area within Germany.

After allocating the number of snow days to each pixel the error mask was added. The corrected seasonal snow maps (modelled and observed) of snow days are shown in Figure 3.7.

3.5.3 Predicting future snow cover days

Once the networks have been trained we used the meteorological time series of Orlowsky et al (2007) as input variables to the NARX model to generate snow time series for the time slots 2021-2030 and 2041-2050. Both scenario realizations MF (wet) and MT (dry) have been evaluated.

Figure 3.8 and 3.9 shows the changes in snow days at the Feldberg station (see Appendix D for the other stations). The black dots represent the mean number of snow days computed for the validation period while the white dots represent the mean number of snow days for the appropriate scenario MF or MT. Absolute changes are indicated by the filled rectangles (decrease) and blank rectangles (increase). The 90% prediction bounds (black lines) for each month are calculated by means of the validation simulations assuming a centred and reduced normal distribution around the estimator. If changes are greater than the derived

3.5 Results

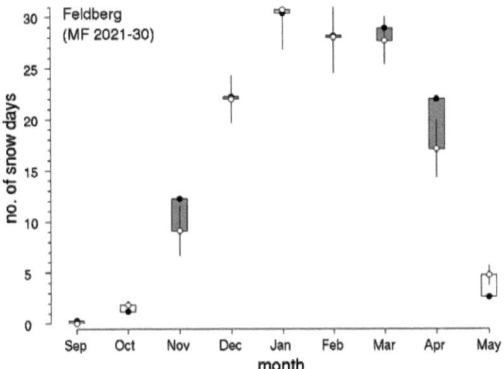

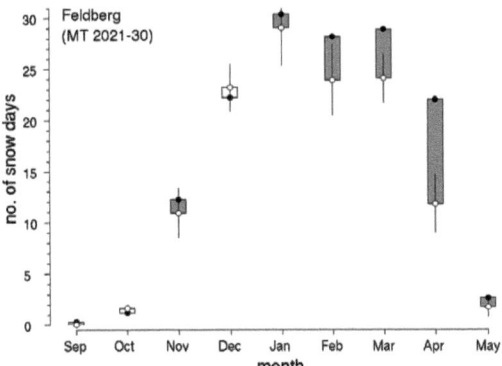

Figure 3.8: Monthly snow-cover days' changes at the location Feldberg by 2021-30 for different scenarios (MF = wet, MT = dry; for a detailed description of the scenario data, see Section 3.3. The plot is described and discussed in Section 3.5.

3 Spatio-temporal prediction of snow cover in the Black Forest mountain range

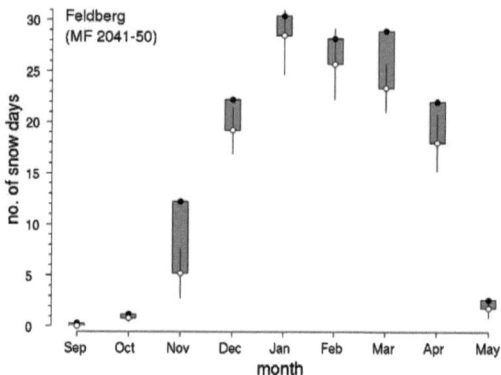

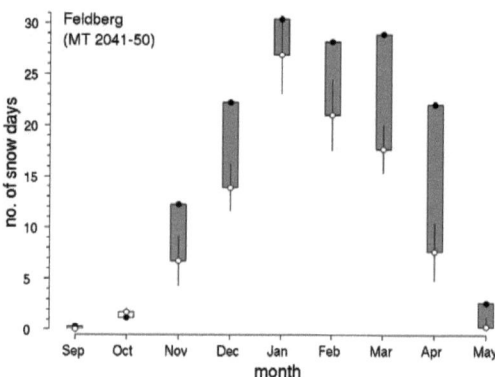

Figure 3.9: Monthly snow-cover days' changes at the location Feldberg by 2041-50 for different scenarios (MF = wet, MT = dry; for a detailed description of the scenario data, see Section 3.3. The plot is described and discussed in Section 3.5.

Table 3.4: Mean relative changes (%) and absolute (days) changes in snow days for different decades and scenarios with varying moisture content.

Decade	Elevation interval					
	> 1200 m		1001-1200 m		501-1000 m	
2021-2030 (wet)	-0.2	0 days	-9.01	6 days	-18.5	6 days
2021-2030 (dry)	-24.7	21 days	-41.4	26 days	-55.5	18 days
2041-2050 (wet)	-10.4	9 days	-17.4	11 days	-24.9	8 days
2041-2050 (dry)	-43.9	37 days	-57.2	36 days	-65.7	20 days

prediction bounds, the changes in snow cover are significant. According to the model output we may expect a shortening in the duration of the snow season due to an earlier begin of the melting season (Figure 3.8 and 3.9). At the same time there are no significant changes in autumn. This pattern is consistent at all the stations.

With regard to the 'wet' scenario, stations with high elevations of more than 1,250 m a.s.l. are likely to gain more snow days until the period between 2021-2030 due to an increase in winter precipitation (Figure 3.10, 3.11 as well as Table 3.4). In case of the Feldberg (1,493 m a.s.l.) the number of snow days will increase by more than 10%. Overall, the number of snow days will remain stable at high elevations of more than 1,200 m a.s.l. in the 'wet' scenario. However, at lower elevations, the number of snow days will decrease by approximately 20% at elevations ranging from 500 m to 1,000 m a.s.l. and almost 10% within a range of 1,000 m to 1,200 m a.s.l. Overall, at all locations above 500 m a.s.l the number of snow days will decrease by almost 10%.

Considering the 'dry' scenario a downward trend can be observed at all altitudes in the period from 2021 to 2030. While the reduction in the number of days with snow will remain between 0 and 24% at altitudes of more than 1,200 m a.s.l. the downward trend will intensify dramatically at lower elevations. Regions between 1,000 and 1,200 m a.s.l will have to cope with a loss of snow days of almost 41%, and lower regions between 500 and 1,000 m a.s.l. of approximately 55%. Overall, the number of days with snow will decrease by 40%. In comparison with the 'wet' scenarios these higher losses are based on lower winter precipitation.

With regard to the period between 2041 and 2050 all downward trends of snow days will increase significantly at all altitudes in both scenarios (Figure 3.12, 3.13 as well as Table 3.4). In principle, the patterns between the dry and the wet scenarios are similar to those already described. The decreasing number of snow days is considerably higher at low elevations while this decrease becomes smaller with increasing elevation. Overall, the decrease for elevations ranging from 500 m to 1,000 m a.s.l. is almost 25%. For sites between 1,000 m and 1,200 m a.s.l it is 17% and for sites of more than 1,200 m a.s.l. it is 10% for the 'wet' scenario. As already described for the time period from 2021 to 2030 the pattern of decreasing snow days is even more dramatic for the 'dry' scenario. The number of snow days will decrease by 66% at low elevations (500-1,000 m a.s.l.), by approximately 57% at middle elevations (1,000-1,200 m a.s.l.) and by nearly 45% at high elevations (> 1,200 m a.s.l.).

3.6 Summary and conclusions

A new approach predicting snow depth and distribution using the NARX model has been presented in this paper. The model is driven by meteorological data obtained from regional climate simulations based on the A1B scenario. By means of the generated snow time series and MODIS satellite data we created high resolution snow cover maps (500 m) for the Black Forest Mountain Range (Germany) for the periods of 2021-2030 and 2041-2050. These maps are intended to provide a decision-making tool for local ski-lift operators in order to evaluate future investments.

The nonlinear approach based on NARX reproduced the statistical characteristics of the non-stationary snow time series. Differences in the first and second order moment statistics between the original and modelled time series are small at all stations. Major deviations in the distributions of snow depths appear predominantly in times of high values of snow depths. However, we proved that the underestimates are not systematic but rather that they originate from a random stochastic process. This likewise might be a lack of further influencing variables. Since each station was modelled separately we also found weaker correlations between stations. To preserve these correlations all time series should be trained simultaneously which leads to a drastic increase in calibration time. In order to minimize the cost of calibration it is preferable to train each station separately.

Depending on the scenario selected the model shows a decrease in snow cover at elevations relevant for skiing (above 1,000 m) of approximately 10-30% for the time period of 2021-2030. Eventually, snow cover might be reduced by up to 66% by 2041-50 at corresponding elevations. Compared with earlier studies that focused on accurate snow prediction models in the Black Forest (Schneider et al., 2007a, 2005), the new approach has considerably improved the prediction of spatio-temporal patterns of snow. It would be interesting to use the results of this study for other applications such as hydrological run-off modelling in the future.

3.6 Summary and conclusions

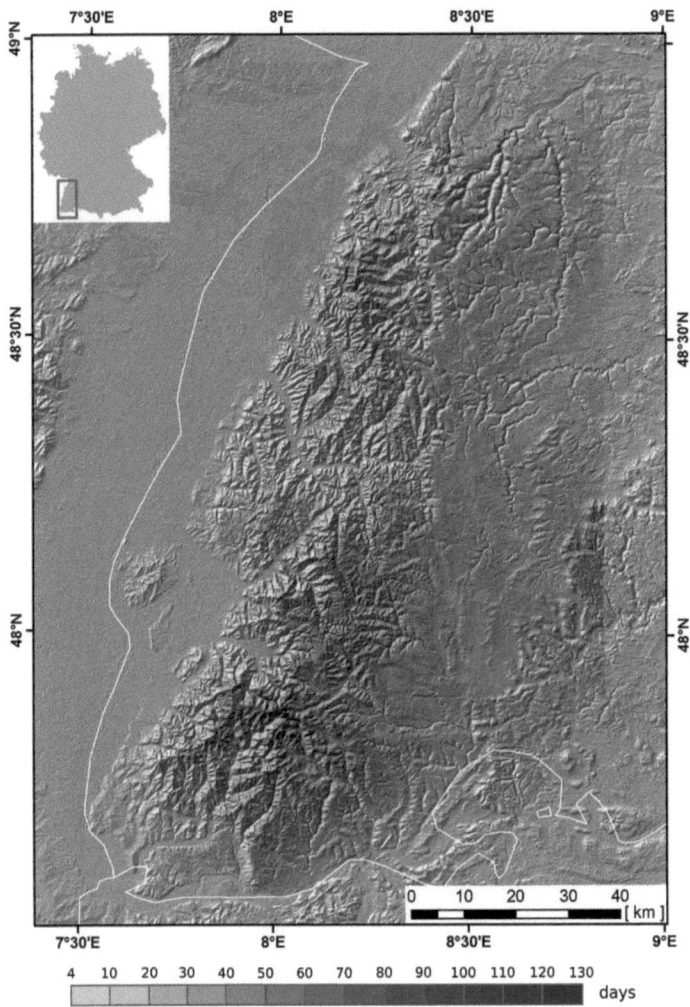

Figure 3.10: Snow-cover maps for the MF scenario for the decade 2021-2030; (for a detailed description of the scenario data, see Section 3.3). Only heights above 250 m are considered. The inset shows the location of the study area within Germany.

3 Spatio-temporal prediction of snow cover in the Black Forest mountain range

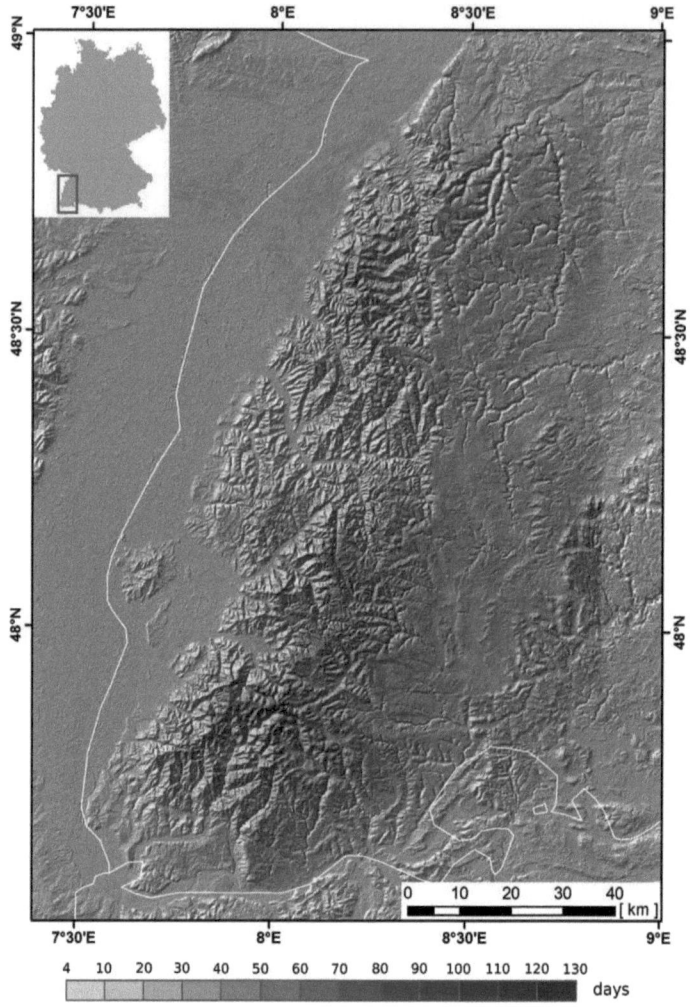

Figure 3.11: Snow-cover maps for the MT scenario for the decade 2021-2030; (for a detailed description of the scenario data, see Section 3.3). Only heights above 250 m are considered. The inset shows the location of the study area within Germany.

3.6 Summary and conclusions

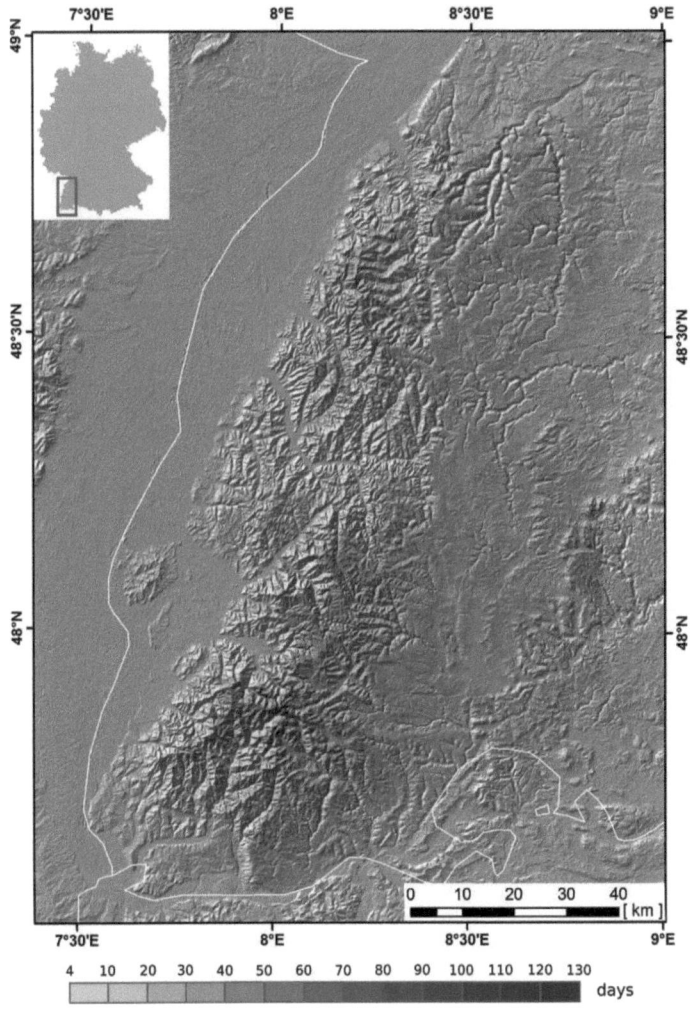

Figure 3.12: Snow-cover maps for the MF scenario for the decade 2041-2050; (for a detailed description of the scenario data, see Section 3.3). Only heights above 250 m are considered. The inset shows the location of the study area within Germany.

3 Spatio-temporal prediction of snow cover in the Black Forest mountain range

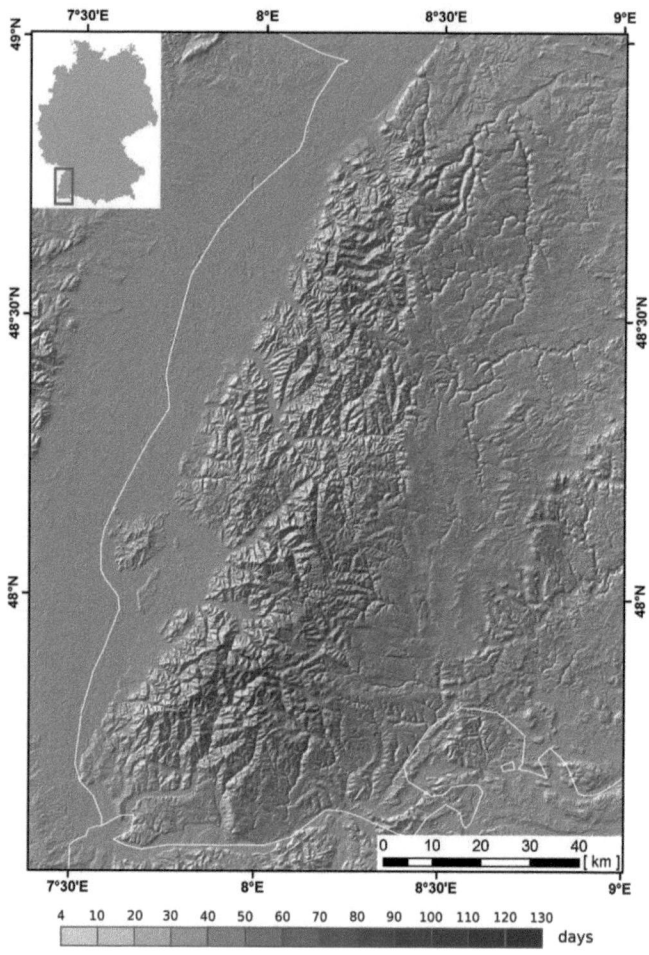

Figure 3.13: Snow-cover maps for the MT scenario for the decade 2041-2050; (for a detailed description of the scenario data, see Section 3.3). Only heights above 250 m are considered. The inset shows the location of the study area within Germany.

Chapter 4

Natural three-dimensional predictor domains for statistical precipitation downscaling

---

The question of whether optimized predictor domains and choice of predictors improve the predictive power of statistical precipitation downscaling is addressed. In addition the sources of predictor uncertainty for precipitation downscaling on daily scale are studied, as well as the skills of the potential atmospheric predictors. With this in mind, we pursue a two-fold approach: First, predictors and their corresponding 3D-domains are optimized by a conditional air mass classification. Since no restrictions are made on the shapes of the domains, the method allows for irregular spatial boundaries. Furthermore, in this way the air masses could be optimized to have a wide range of different precipitation distributions. Second, an artificial neural network as a non-parametric statistical downscaling model is applied in order to estimate the sensitivity of individual predictors and their interactions. Inferences on the sources of uncertainty are drawn from model-free sensitivity measures also permitting to capture interaction effects. Using the optimized predictors improves the accuracy of the downscaled time series, in particular in seasons with strong convection. According to a global sensitivity analysis geopotential height, vertical velocity, temperature and specific humidity are the most influencing predictors for precipitation downscaling.

4.1 Introduction

Statistical downscaling has become a well-established tool in regional and local impact assessment over the past few years. Especially precipiation downscaling is paramount for impact studies to correct its spatial and temporal structure. In order to resolve the scale discrepancy between Global Circulation Models (GCM) and local-scale weather, a vast number of algorithms and methods have been proposed and also extensively tested. A comprehensive review on different precipitation downscaling approaches and their advantages and drawbacks is given by Maraun et al. (2010). Despite the extensive research on downscaling methods there is still little consensus about the choice of useful atmospheric predictor variables (Wilby and Wigley, 2000). Besides the general decision of a proper statistical downscaling model, the selection of an *informative predictor* set is crucial for the accuracy and stability of the resulting downscaled time series (Fealy and Sweeney, 2007; Huth, 2004; Fowler et al., 2007; Maraun et al., 2010).

This statement relates both to the atmospheric variables as well as the predictor domains in terms of geographical location and spatial extend, to which in general not much attention is paid, which was also noted by (Goodess and Palutikof, 1998; Wilby and Wigley, 2000). Ideally, suitable predictors for perfect-prog (PP) approaches should fulfill the following conditions: (i) stationary relationship between predictors and predictands, (ii) valid relationship under future climate conditions, (iii) predictors should have a high predictive power (explained variance) and (iv) predictors should be reasonably well simulated by the driving dynamical model (Maraun et al., 2010). For the first two conditions it is important that the statistical relationships found can be understood in physical terms.

Some circulation-related predictors, such as sea-level pressure, fulfill most requirements and are therefore often employed as predictors in PP-downscaling studies. However, on small scales, the sole use of circulation predictors often turns out to be insufficient to capture the complete variability. Therefore some studies additionally consider thermodynamic variables and the amount of available water in the atmosphere (e.g. Wilby and Wigley, 1997; Murphy, 2000; Beckmann and Buishand, 2002). Cavazos (2005) assessed the relative skill and error of 29 potential atmospheric predictors from the NCEP/NCAR reanalysis and their relevance for precipitation at 15 globally distributed locations. An Artificial Neural Network (ANN) downscaling model was applied, including most relevant predictors of daily precipitation. Then, the most relevant predictors were determined by a rotated principal component analysis. Mid-tropospheric geopotential heights and mid-tropospheric specific humidity were found to be the dominant predictors at all locations. Most other predictors, however, show a strong regional and seasonal dependence.

Common classification approaches usually operate on both a fixed regular domain such as the NCEP/NCAR reanalysis data and a predefined number of predictors. This approach is convenient for many traditional large-scale weather type classification methods. However, for finding linkages between atmospheric characteristics and local phenomena this may not be optimal. In fact, one would expect atmospheric subspaces with natural shapes varying in time and space (e.g. Huth, 1999; Philipp et al., 2007).

Only a limited number of papers is interested in the predictive capability of the domain size or shape and the question to what extent variablility of neighbouring lattices influence local-scale events. Wilby and Wigley (2000) emphasized the spatial relationships between observed daily precipitation and both mean sea-level pressure and atmospheric moisture for selected regions in the United States. They found that in many cases, maximum correlation between precipitation and mean sea-level pressure is not directly located above the target grid cell, whereas correlations with specific humidity were largest when the predictors were right above the predictant position. Recently, D'onofrio et al. (2010) used a weather pattern classification to optimize the spatial domain of predictors using a set of 17 atmospheric variables. They found that the optimum spatial domain is smaller than the synoptic scale and therefore suggested to consider different domains for dynamical and non-dynamical predictors.

Instead of calculating correlation measures to identify potential predictors and domains, we make use of the local precipitation distribution by applying a conditional air mass classification. The proposed approach aims to optimize the choice of predictors and their corresponding input domain used for conditional classification (see Figure 4.1). Unlike common classification methods, we optimize the input to a semi-objective cluster algorithm to extract atmospheric states with similar observed local scale precipitation distributions and allow irregular spatial boundaries for each predictor domain. The optimization requires a cost-function specifying the desired criteria for clustering, in this case significantly different precipitation distributions of each class. Once, a proper criteria is defined, the classification becomes an optimization problem in terms of minimizing the user-defined cost-function. The approach is not constrained to a specific clustering or optimization algorithm; for this study a nonlinear unsupervised clustering method called Self-Organizing Maps (SOM) is utilized, while a generic and robust probabilistic Simulated-Annealing (SA) algorithm is used for optimization.

So far no downscaling study paid much attention to either the interactions of individual variables or grid lattices, nor to the different sources of uncertainty in the model. If one would like to achieve a full understanding of the model's sensitivity pattern it is necessary to recover the complete variance of the predictand, especially for nonlinear non-additive models (Saltelli et al., 2006). Hence, a model-free sensitivity measure is required that is valid independently of the degree of linearity. An attractive model-free approach are variance-based methods, which can also account for nonlinear non-additive models, unlike methods based on normalized derivatives.

To analyse the sensitivity of individual predictors we need a robust statistical model describing the relationship between the local scale predictant and the previously optimized predictors that is able to account for nonlinear empirical relationships in the observed system. Thus we utilise ANNs, just as many other downscaling studies (e.g. Cavazos, 2005; Hewitson and Crane, 1996; Schoof and Pryor, 2001; Wilby et al., 1998; Coulibaly et al., 2005).

4 Natural three-dimensional predictor domains for statistical precipitation downscaling

Once, a robust model is trained, the first- and higher-order effects are estimated by Monte-Carlo (MC) based Global Sensitivity Analysis (GSA). Taking advantage of the MC runs we also analyse the model output within specific bounds of interest (*factor mapping*). The proposed algorithm has been applied and evaluated to daily precipitation data in the Rhineland region of Germany.

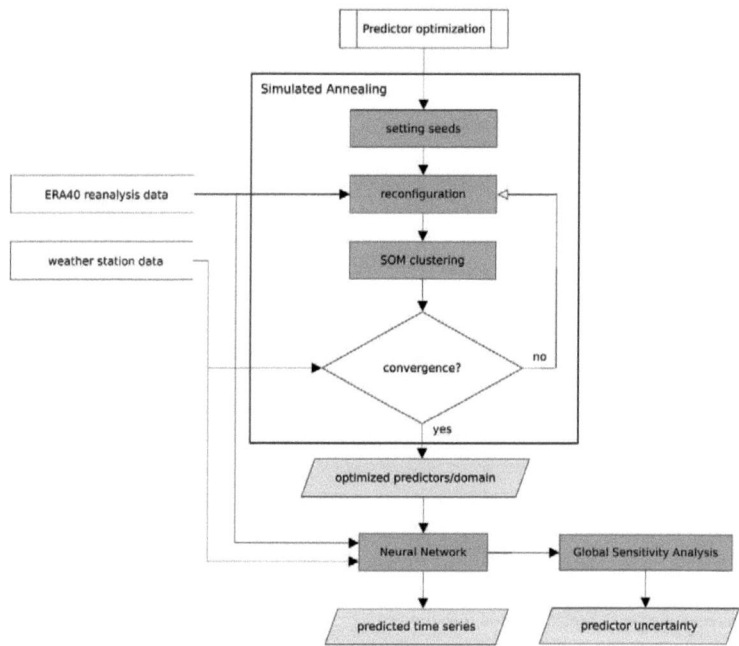

Figure 4.1: Flowchart of the proposed predictor optimization process.

4.2 Predictor selection

4.2.1 Self-organizing maps

Self-organizing maps is one of a multiplicity of unsupervised clustering algorithms and has become more popular, also in the field of atmospheric science, during the last decades (Crane and Hewitson, 2003; Hewitson and Crane, 2002; Lynch et al., 2006; Kohonen, 1982). Due to their structural similarity to Artificial Neural Networks, SOMs are considered as a special case of ordinary ANN. Rather than identifying characteristic regions in data space by maximizing the within-group similarity and the difference between groups, SOMs aim to find a predefined number of representative multi-dimensional prototypes (position vectors) in data space. These prototypes are mapped onto a lower dimensional map space, which is intended to preserve the topological properties of the input space. This low-dimensional presentation is, in particular, useful in describing and visualizing high-dimensional data. In the beginning of the learning phase prototypes are initialized randomly in data space and are adjusted iteratively during the learning process.

Let M be a set with μ_M input data points $\mathbf{m_i}$ each specified by a n-dimensional data vector $\mathbf{x_i}$

$$M = \{\mathbf{m_i} = (\mathbf{x_i}) | \mathbf{x_i} \in X \subseteq \mathbb{R}^n, i = 1, ..., \mu_M\} \tag{4.1}$$

So we can define another set N consisting of neurons (prototypes) composed of a weight vector $\mathbf{w_i}$ within the same input data space and a node vector $\mathbf{k_i}$ located on a discrete two dimensional topological map $\mathbb{K}^2$ (Kohonen-map).

$$N = \{\mathbf{n_i} = (\mathbf{w_i}, \mathbf{k_i}) | \mathbf{w_i} \in X \subseteq \mathbb{R}^n, \mathbf{k_i} \in \mathbb{K}^2, i = 1, ..., \mu_N\} \tag{4.2}$$

In general higher- or lower dimensional topological maps are possible, but not considered in this work. At each iteration step t we randomly select one of the data vectors $\mathbf{m}_j^t$ and calculate the metric distance to each prototype $\mathbf{w}_i^t$ in X.

$$\| \mathbf{x}_j^t - \mathbf{w}_w^t \| = \min\{\| \mathbf{x}_j^t - \mathbf{w}_i^t \| \, | i = 1, ..., \mu_N\} \tag{4.3}$$

The neuron n_w with the least distance is called the 'Best Matching Unit' (BMU). Subsequently, the weights of the BMU and neurons, closer than a certain time-dependent distance threshold δ^t from the winner node on the topological map, will be adjusted. The prototype subset, which will be updated, is defined as

$$N^t = \{n_i = (w_i, k_i) | \; \| k_w - k_i \| \leq \delta^t, i = 1, ..., \mu_N\} \qquad (4.4)$$

All vectors of the subset N^t are slightly adjusted towards the selected data point x_j^t according to Hebb's learning rule:

$$w_i^{t+1} = w_i^t + \varepsilon^t + h^t + \| x_j^t - w_i^t \| \qquad (4.5)$$

The degree of adaptation is determined by the learning rate ε^t, the weight function h^t and the distance $\| x_j^t - w_i^t \|$ between the prototype and the data vector. During the learning process the learning rate smoothly decreases by

$$\varepsilon^t = \varepsilon_{start} \cdot \left(\frac{\varepsilon_{end}}{\varepsilon_{start}}\right)^{\frac{t}{t_{max}}} \qquad (4.6)$$

whereas ε_{start} and ε_{end} are the learning rates at the beginning and the end of the procedure, respectively. The weight function (Gaussian update kernel) is reduced by a time-dependent adaptation radius

$$\delta^t = \delta_{start} \cdot \left(\frac{\delta_{end}}{\delta_{start}}\right)^{\frac{t}{t_{max}}} \qquad (4.7)$$

Once, no more changes in location of prototypes are observed, or the maximum number of iterations is reached, the algorithm terminates. Under perfect conditions prototypes are arranged such that they span the entire data space along with more prototypes in regions of high data density (Hewitson and Crane, 2002; Kohonen, 1982). The properties of SOMs allow to approximate a nonlinear distribution in data space despite the linear measures. However, in order to run SOMs the user must define a large number of parameters such as the size and shape of the update kernel, learning rates and number of prototypes. After training each data point is assigned to the closest node.

4.2.2 Simulated Annealing

The problem finding an optimal predictor subset on a discrete configuration space is an example of combinatorial minimization. Solving this problem with an exhaustive search algorithms requires exponential computing time; this problem thus belongs to the complexity class 'NP-complete'. For a large number of grid points in the configuration space, it is impossible to explore all possible combinations. In the case study presented in this paper the number of possible solutions is 2^{11730}. In most cases such combinatorial problems can be solved, or at least well approximated, by the heuristic probabilistic Simulated Annealing

4.2 Predictor selection

(SA) algorithm (Kirkpatrick et al., 1983). Analog to thermodynamic processes the SA algorithm simulates the slow cooling (annealing) a hot of crystal. During the cooling the thermal mobility of atoms gradually reduce until the system reaches a state of minimum energy. As the energy distribution of a physical system in thermal equilibrium is given by the Boltzmann probability distribution, a system at low temperature has also a chance of being in high energy state. This distribution increases the chance for the system to get out of a local energy minimum. For the algorithm, this means that changes that lead to improvements of the cost function are always implemented and the changes that increase the costs are implemented with a finite probability to be able to avoid local minima. The concept of simulated annealing was first introduced in numerical calculations by Metropolis et al. (1953) and has since become a widely applicable heuristic optimization technique.

The iterative algorithm starts with a predefined domain configuration by setting 'seeds' for each predictor. In each step, the configuration is rearranged by adding or removing one randomly selected grid point, after which the change in energy ΔE (cost-function; see below) between the old and new configuration is computed. Based on ΔE the system changes its configuration from energy E_1 to energy E_2 with probability

$$\Pr(\Delta E) = exp\left(\frac{-\Delta E}{T}\right) \quad (4.8)$$

where T is a control parameter (analog of temperature). If $\Delta E \leq 0$, i.e. the solution becomes better, the the rearrangement is always accepted ($\Pr(\Delta E) = 1$). In case $\Delta E > 0$, i.e. the solution becomes worse, the solution is accepted with this finite probability, $\Pr(\Delta E)$, to avoid getting trapped in a local minima in the cost function. During optimization, T is gradually lowered (see Table 4.1). The SA algorithm was implemented with the following modules.

Rearrangements. At the beginning of each iteration step the algorithm determines whether a new grid point is added or a seeded grid point is removed. The ratio of additions to removals varies in time and oscillates harmonically with a frequency of 0.001 between a ratio of 0.7 and 0.3. This strategy reduces the chance that the algorithm gets stuck in a local optimum. If grid points are added, the chance of selecting one of the directly surrounding grid points is higher (0.9) than selecting farther grid points. In case no improvement is observed after 2000 consecutive iterations the configuration is set back to the best configuration.

Cost-function (Energy). The classification is conditioned on the local scale precipitation distributions measured at eight weather stations. For every possible pair of classes the χ^2 test is used to determine whether the two distributions significantly differ on a 95%-level. The optimization algorithm maximizes the total number of different distributions. Additionally, the χ^2 values are summed up to get a measure how different distributions are. In case two model runs have the same number of different distributions, the final decision is determined by the summed χ^2 value.

4 Natural three-dimensional predictor domains for statistical precipitation downscaling

Table 4.1: SA paramter set.

Description	Value
maximum number of iterations at each temperature	2000
maximum number of successful reconfigurations	100
cooling rate	10%
convergence criteria (temperature T <convergence)	10^{-4}
no. of iterations before set back to best configuration	400
ratio between adding or removing grid points	0.3-0.7
frequency of ratio of adding or removing grid points	0.001

Annealing schedule. The annealing schedule is determined empirically. The temperature is hold constant for 1000 reconfigurations, or for 100 successful reconfigurations in order to reach approximately thermal equilibrium. Consecutively the temperature is lowered by a constant cooling rate of 10%. Due to the large number of grid points this rate does not guarantee finding the global minimum.

4.3 Global Sensivity Analysis (GSA)

To investigate the nonlinear and non-additive effects of the predictors, at first ANN models are trained to approximate the functional link between local scale precipitation and the SOM optimized predictors. Subsequently, these effects are estimated by decomposing the output variance of these models by a GSA. In general, sensitivity analysis permits inferences on the different sources of uncertainty in the model input by decomposing the variance of the model output. For the sake of clarity this section re-iterate how model-free sensitivity measures are derived from variance-based methods, as already mentioned in Section 2.2.2. Let us assume a generic model

$$Y = f(X_1, X_2, \cdots, X_k) \qquad (4.9)$$

with the corresponding total or unconditional variance $V(Y)$. If one factor (predictor) X_i is fixed at a particular value x_i^*, the resulting conditional variance of Y is accondingly $V_{X\sim i}(Y|X_i = x_i^*)$. Since the conditional variance will be less than the unconditional variance this measure characterises the relative importance of the factor X_i. The fact that, this sensitivity measure depend on the value of x_i^* makes it rather impractical. Taking instead the average of this measure over the valid ranges of x_i^* with uniform distribution, the undesired dependence will disappear (Saltelli et al., 1999, 2006). We can obtain following expression

$$E_{X_i}(V_{X\sim i}(Y|X_i = x_i^*)) + V_{X_i}(E_{X\sim i}(Y|X_i = x_i^*)) = V(Y) \qquad (4.10)$$

where the second conditional variance on the left hand side is called the first-order effect of X_i on Y. The corresponding first-order sensitivity index of X_i is given by

$$S_i = \frac{V_{X_i}(E_{X_{\sim i}}(Y|X_i = x_i^*))}{V(Y)} \qquad (4.11)$$

This sensitivity index indicates the importance of individual factors without considering interactions effects. In case the model belongs to the class of additive models, the first-order terms add up to one, e.g. $\sum_{i=1}^{r} S_i = 1$. If this is not the case, the variance must be explained by the higher-order effects (interaction) between factors. Interactions represent an important feature, especially, of nonlinear non-additive models. The total sensitivity S_{T_i} of a factor X_i is made up of the first- and all higher order terms where a given factor X_i is participating, consequently giving information on the nonadditive character of the model. The S_{T_i} can be computed using,

$$S_{T_i} = \frac{S_{Y|X_i} + S_{Y|X_{i,\sim i}}}{S_Y} \qquad (4.12)$$

where $S_{Y|X_{i,\sim i}}$ indicates all importance measures involving the factor X_i. This approach permits, even for nonadditive models, to recover the complete variance of Y. The indices can be efficiently computed by Monte-Carlo based numerical procedures (Saltelli et al., 2006, 1999).

4.4 Case study: Rhineland region

4.4.1 Data and set-up

Daily precipitation measurements for 1971 to 2000 have been obtained from the German Weather Service (DWD) for 8 weather stations in the Rhineland region (Germany) providing a set of 8x10592 samples (Figure 4.2). For this period complete data sets are available for all stations. The large-scale atmospheric data is derived from the ECMWF reanalysis ERA-40 provided in 2.5° x 2.5° regular geographical coordinates. The potential predictor domain is centered on Central Europe ranging from 35°N-75°N and 12.5°W-42.5°E. We used the variables: temperature (T), specific humidity (sH), vertical velocity (w), divergence (div), potential vorticity (pv) and geopotential height (z) on 5 geopotential levels (1000, 850, 700, 500 and 300 hPa) for classification. Inherently, some of the used predictor variables show significant correlations, e.g. sH and T, or div and w.

The algorithm was initialized by setting the seed for each predictor to the grid point closest to the precipitation observation (5°E, 50°N) in the potential predictor domain. On the

4 Natural three-dimensional predictor domains for statistical precipitation downscaling

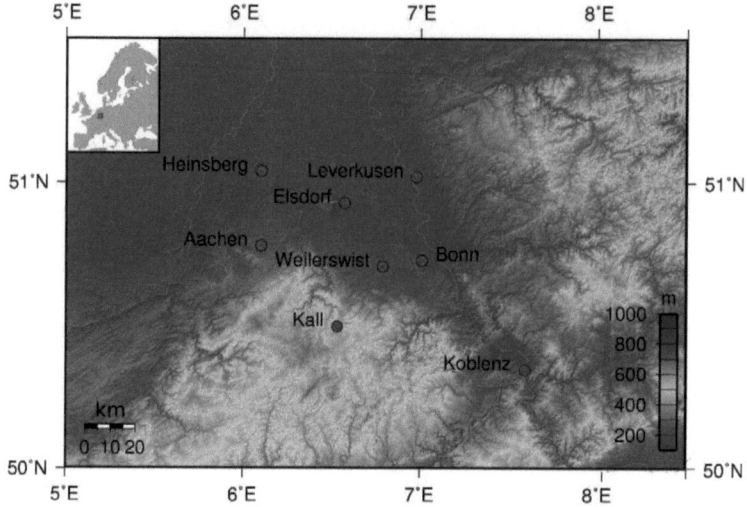

Figure 4.2: Digital Elevation Model of the area of investigation with marked location of the weather stations. The inset shows the location of the study area within Europe.

High Performance Clusters (HPC) of the University of Aachen several computations with a range of SOM node configurations have been performed. From the set of models with different precipitation distributions for all classes, the configuration with the largest number of precipitation classes was chosen to be the best model. According to this criterion the best SOM architecture consists of 6 nodes (see Figure 4.3). To assure that the algorithm has converged we restarted the optimal configuration and compared the two runs. The second run converged to the same subdomains indicating that the algorithm found a robust optimum.

4.4.2 Predictor domains

Figures 4.4 and 4.5 show the optimized subdomains used for the final classification. Note that the algorithm identifies both continuous and disjointed subdomains (e.g. T, pv). In the following predictor domains are analysed concerning their spatial structure as well as by their inherent dynamic and thermodynamic characteristic.

The atmospheric moisture content sH subdomain is situated directly above the study area and its spatial extent is much smaller than those of the other subdomains. Wilby and Wigley (2000) found similar spatial correlations patterns between precipitation and sH for six regions in the USA, tending to overlap with the target. Together, with the w pattern, extend-

4.4 Case study: Rhineland region

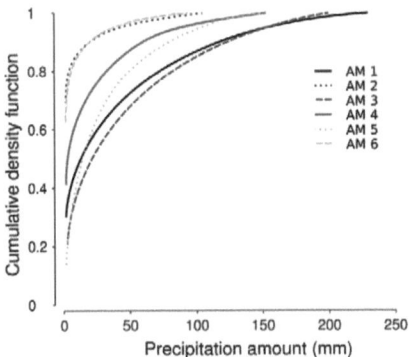

Figure 4.3: Cumulative densitiy distribution of precipitation amounts for each of the six airmasses at the weather station Aachen.

ing upward to a height of several kilometers (300 hPa) and sloping backward toward the west, this formation capture air masses advancing aloft a frontal surface. This pattern is well-known from rising warm air advection in advance of developing surface lows and condensation of water vapor in the upper atmosphere.

The *div* dipole is situated between the upper (300 hPa) and ground-level atmosphere (1000-700 hPa) with the lower region directly connected to the lower edge of the *w* subdomain. This is essentially due to the close linkage of the two predictors by the continuity equation. Depending on whether air masses are ascending or descending (in the *w* domain) the two *div* regions behave either as sinks or sources (see Table 4.2).

Useful information on diabatic processes can be obtained by changes in the *pv*, (Davis and Emanuel, 1991; Rossa et al., 2000). As *pv* is the product of absolute vorticity on an isentropic surface and static stability, a release of latent heat is concentrated within the domain of enhanced *pv* values. Therefore, strong baroclinic zones, where much diabatic heat release takes place, usually form low level *pv*-anomalies (near the ground, see also Figure 4.4). Besides the derivation of diabatic processes the vorticity equation admits, by mean of its invertibility, inferences on other meteorological fields such as geopotential, temperature, wind and static stability. In the North Atlantic region, the highest cyclone track densities are found around 60°N (Ulbrich et al., 2008, 2009; Wang et al., 2006), slightly varying with the seasons. Consequently, significant changes of *pv* particularly take place in this zone. This is in good agreement with the obtained *pv* subdomain, which ranges from the northern part of Germany along the west coast of Norway up to 62.5°N. The structure also nicely reflects the often observed comma-shaped forming in the proximity of low pressure systems.

4 Natural three-dimensional predictor domains for statistical precipitation downscaling

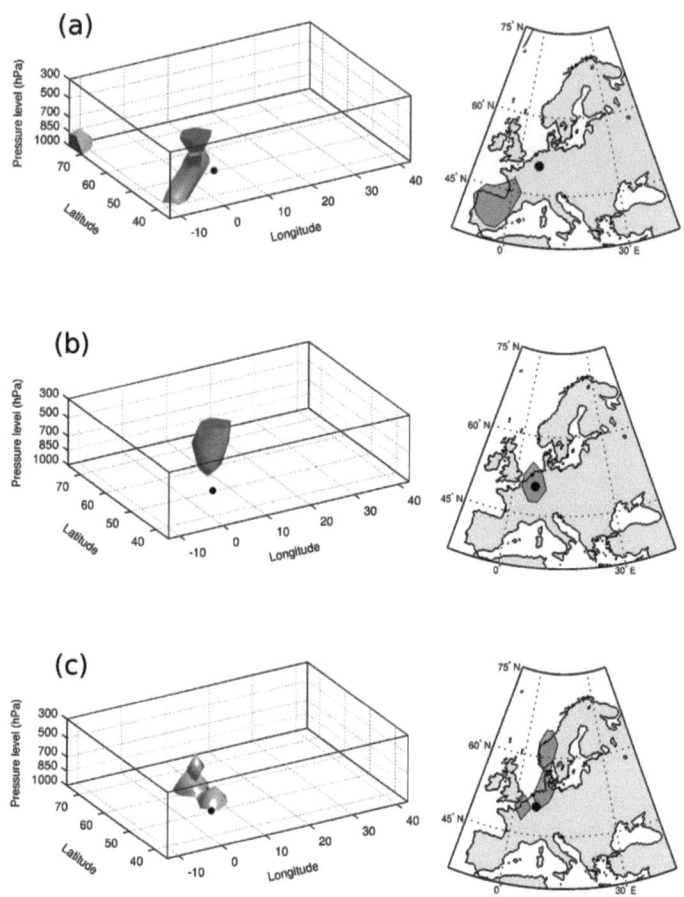

Figure 4.4: Optimized predictor domains for a) temperature, b) specific humidity and c) potential vorticity. The left panels shows the three-dimensional perspectives of the domains. For presentation the lattices are filter by an 3x3x3 gaussian filter and displayed by an isosurface. Also shown are the corresponding top view perspectives (right panel). The black dot indicates the location of the study area.

4.4 Case study: Rhineland region

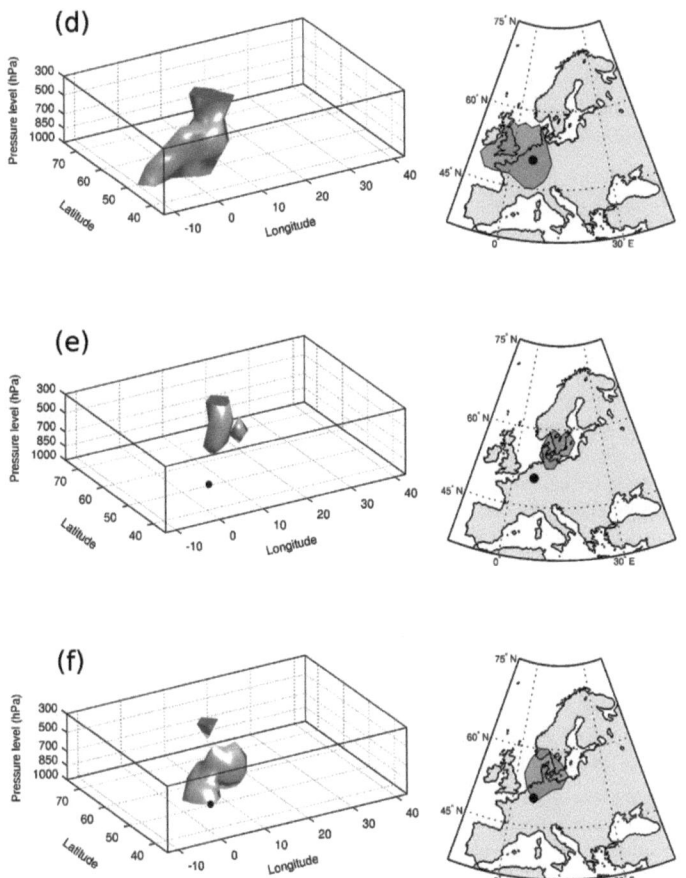

Figure 4.5: Optimized predictor domains for d) vertical velocity, e) geopotential height and f) divergence. The left panels shows the three-dimensional perspectives of the domains. For presentation the lattices are filter by an 3x3x3 gaussian filter and displayed by an isosurface. Also shown are the corresponding top view perspectives (right panel). The black dot indicates the location of the study area.

4 Natural three-dimensional predictor domains for statistical precipitation downscaling

Unlike the subdomains of the circulation and moisture predictors, the two disjointed T domains are located over Spain and Iceland. This is surprising as the formation of precipitation by lifting and cooling of air masses is strongly bound to temperature changes. It is interesting that the location of the two domains show similarites with the North Atlantic Oscillation (NAO) pattern, which is known to influence the amount of winter precipitation. Another surprising result is that the geopotential height does not show the NAO dipole pattern. However, the z domain is shifted northeast to the Rhineland and stretches vertically from the upper to the middle troposphere with small horizontal extent.

4.4.3 Air masses

As indicated before, the six classes are intended to describe distinct air masses rather than synoptic circulation patterns (Grosswetterlagen). Some mean characteristics of the predictants for each class are given in Table 4.2 and the resulting composite mean sea level pressure and specific humidity maps are presented in Figure 4.6, 4.7 and 4.8. These classes can be roughly divided in three low (AM1, AM3, AM6) and three high pressure systems (AM2, AM4, AM6). On basis of their precipitation distributions, the classes can also be subdivided into two humid (AM1, AM3), two moderate (AM4, AM5) and two dry (AM2, AM6) air masses (see Figure 4.3 and Table 4.2).

Frequent rainfall events at the study area are associated with advection of humid air masses from west and southwest, e.g. AM1 and AM3 (see Figure 4.3, Figure 4.6, 4.7 and Table 4.2). In both cases the advection is driven by a low pressure system north of the British Isles. Within the complete w subdomain vertical winds are negative, indicating a large-scale ascent of air masses. Once the frontal system passed the study area and the low pressure system is located above Scandinavia uplift rates reduce (AM5) and the flow changes to northwest. Quantitative differences in moisture content of the air masses are likely to be attributed to seasonal effects (compare Figure 4.9). It is worth mentioning that precipitation amount and occurence are not directly related to moisture content (see Table 4.2). Other physical conditions, like vapour pressure deficit, might be useful predictors for this purpose. As assumed seasonal differences are also present in the mean temperature gradient between the T dipole. The much dryer classes AM2 and AM6 show pronounced high pressure systems over Central Europe accompanied with dry air masses in the mid-troposphere. As a consequence of the large-scale subsidence, divergence take positive values in the complete div subdomain. The associated adiabatic heat release leading to changes in the vapour pressure deficit and hence to a reduction of relative humidity. This is evident above all in the reduced precipitation probabilities.

4.4 Case study: Rhineland region

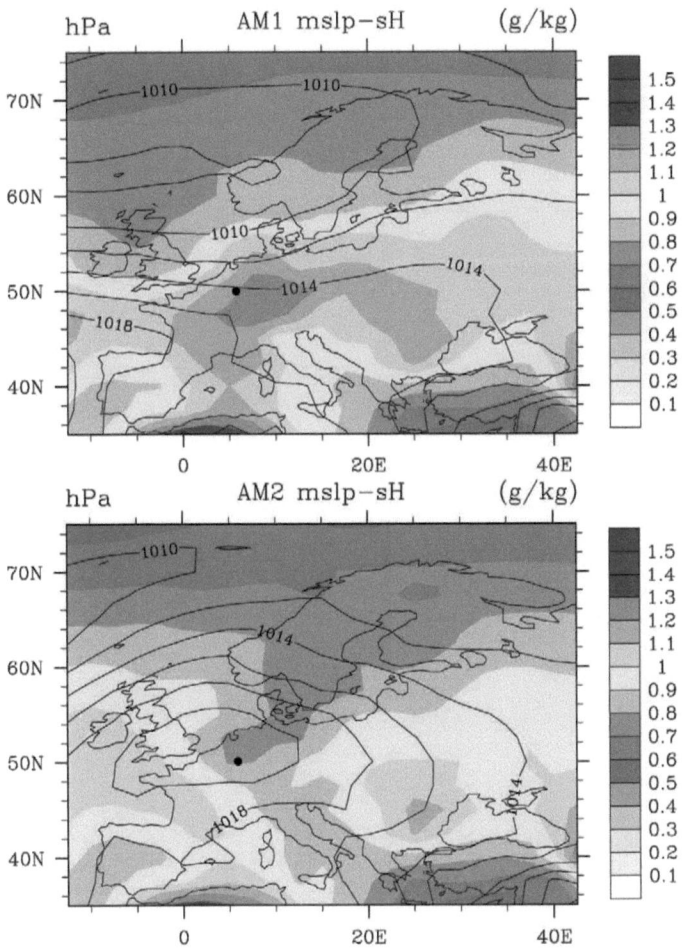

Figure 4.6: Mean sea level pressure fields (contour lines) and specific humidity at the 500 hPa level (filled contour) for AM1 and AM2. The block dot indicates the location of the study area.

73

4 Natural three-dimensional predictor domains for statistical precipitation downscaling

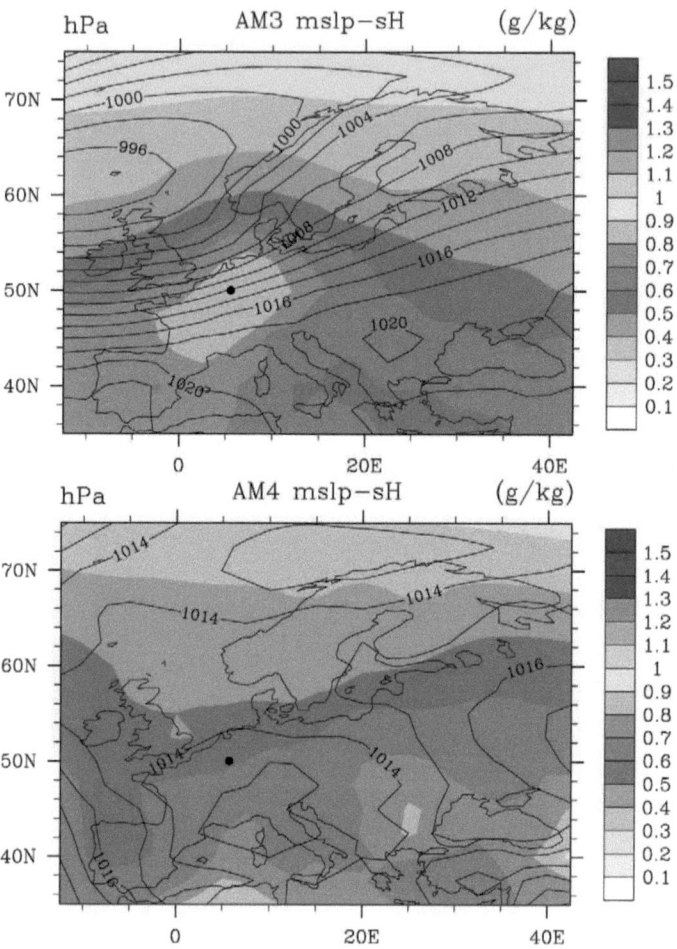

Figure 4.7: Mean sea level pressure fields (contour lines) and specific humidity at the 500 hPa level (filled contour) for AM3 and AM4. The block dot indicates the location of the study area.

4.4 Case study: Rhineland region

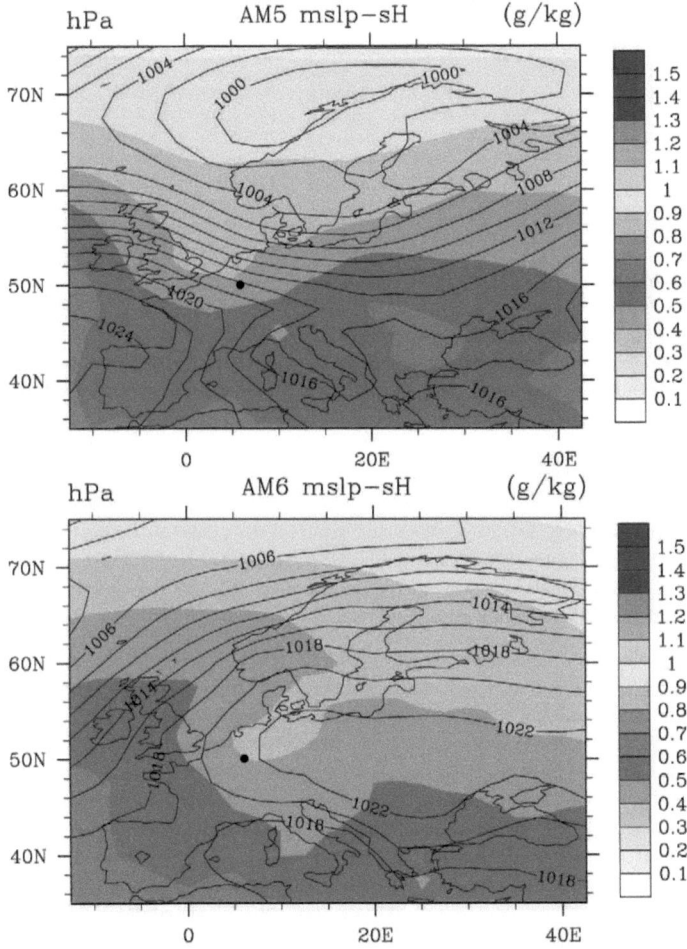

Figure 4.8: Mean sea level pressure fields (contour lines) and specific humidity at the 500 hPa level (filled contour) for AM5 and AM6. The block dot indicates the location of the study area.

4 Natural three-dimensional predictor domains for statistical precipitation downscaling

Table 4.2: Characteristics of air masses for each class. Shown are the mean values of sH and w in the corresponding subdomains. Also given is the mean difference in surface temperature DT and standard deviation (in bracketes) of the T dipole. To gain a detailed understanding also the local mean rain rate RR and fraction of rainy days RD are presented.

Class	$sH\left(\frac{g}{kg}\right)$	$w\left(\frac{Pa}{s}\right)$	div	DT (K)	RR (mm)	RD
AM1	1.5	-0.024	negative	-18.4 (3.5)	3.6	0.68
AM2	0.9	0.029	positive	-19.9 (3.8)	0.6	0.23
AM3	1.0	-0.063	negative	-20.5 (7.1)	4.1	0.77
AM4	0.7	-0.005	positive	-18.5 (5.2)	1.8	0.52
AM5	0.5	-0.007	negative	-21.8 (6.6)	2.4	0.77
AM6	0.5	0.031	positive	-19.4 (7.3)	0.5	0.29

4.4.4 ANN Downscaling

Before the sensitivity of predictors is estimated an empirical relationship between local scale precipitation and the previously optimized predictors on their domains needs to be derived. A simple non-parametric static feed-forward ANN (Model M1) is trained seperately for each season and station, using randomly 70% of the data (see Section 4.4 and 4.4.1) for calibration, 15% for testing (generalization) and the remaining 15% for validation. For comparison a second ANN (Model M2) is trained using only geopotential height (500 hPa level) and specific humidity (700 hPa level) above the target location and its eight neighbouring grid points as predictors. Cavazos (2005) found that these two predictors are the most relevant controls of daily precipitation in the mid-latitudes. The ANNs are trained using 60 different sets of random initial weights in order to avoid local optima. The mean ANN performance over all stations is estimated from all 60 ANN.

Result of both ANN models are shown in Table 4.3. The quality or accuracy is quantified by the explained variance (R^2), root mean squared error ($RMSE$), skill ratio between simulated and observed standard deviation ($Skill$), mean precipitation differences (DP), ratio of the average number of dry spells (no. of DS) with more than three days and the ratio of the average length of these dry spells (length of DS). These ratios are the the number or length, respectively, found in the simulated downscaled data divided by the value of these indices in the measured data. Measures are calculated seperately for each season so that fluctuations in model accuracy are emphasized. In general both models perform better during winter and autumn due to less convective processes in this period. Other studies (e.g. Tolika et al., 2007; Cavazos, 2005; Coulibaly et al., 2005) reported similar annual fluctuations in model performance. While there is not much improvement using the SOM optimized predictor set in spring, significant improvements are achieved in summer, winter and autumn. Apart from this, it is noticable that the mean precipitation amount is underestimated (between 3 − 11%) by almost all models and seasons. The same holds true for the number and length of consecutive dry day indicating difficulties in modelling days without precipitation. Overall, the downscaling with the optimised domains (M1) seems to be slightly better in reproducing the variability of precipitation; see column Skill. Based on the results we assume that the ANN models predict processes sufficiently accurate so that it may be used for prognostic modelling and the global sensitivity analysis in the next section.

4.4.5 Global Sensitivity Analysis

To understand the importance of the predictors the model's complete sensitivity pattern is required (factor prioritization). In the following section, different sources of uncertainty in the ANN are estimated using the variance-based GSA method introduced in Section 4.3. To take into account the spread of ANN network weights (generalization) first- and higher order indices are calculated and averaged over the best 20 ANN (smallest *RMSE*) at each station. Then, for reasons of clarity indices are averaged over all stations, as well.

Figure 4.10 shows the relative importance of individual predictors given by the total- and first-order indices. First-order indices S_i measure individual predictor contributions to the ANN output variance, while the total-order indices S_{Ti} also include all interaction effects. Whereas the first-order sensitivity is often (close to) zero, in all cases S_{Ti} values significantly differ from zero indicating that all variables contribute to the output variance either singularly or in combination with other variables. By subtracting S_i from S_{Ti} one obtains a measure of how much variables are involved in interactions.

The individual variables account for about 73% (sum of first-order indices) of the output variance, thus the remaining 27% is due to interactions (non-additive). Thus the ANN network, trained for precipitation downscaling, belongs to the class of non-additive models. Apparently, some inputs mainly contribute due to their interactions with other variables, e.g. *div* in JJA. Nevertheless, these factors make an important contribution to the model's output variance by interaction. According to the result *z*, *w*, *T* and *sH* may be designated as sensitive variables as the output is mainly influenced by the uncertainty of these factors. There is clear evidence that these factors are subject to seasonal fluctuation. For example, the geopotential height, *z*, is more important in the lower troposphere around the 500 hPa level during summer and spring, whereas the upper troposphere (300 hPa level) gain importance in winter. On the contrary, at the 850-500 hPa level *w* and *T* are giving a consistent contribution to the model output variance. Note that specific humidity, *sH*, at the 500 and 700 hPa level show high values in summer and a minimum in winter.

Downscaling studies often address the question, which predictors are most responsible producing strong precipitation events (factor mapping). To do so model output realizations are mapped back onto the input space (Saltelli et al., 2006) using the previously used Monte-Carlo realizations from the GSA. In a first step the output is filtered into two subset: one which contains the upper 5% tail of rainy days and a second subset with the remaining

4 Natural three-dimensional predictor domains for statistical precipitation downscaling

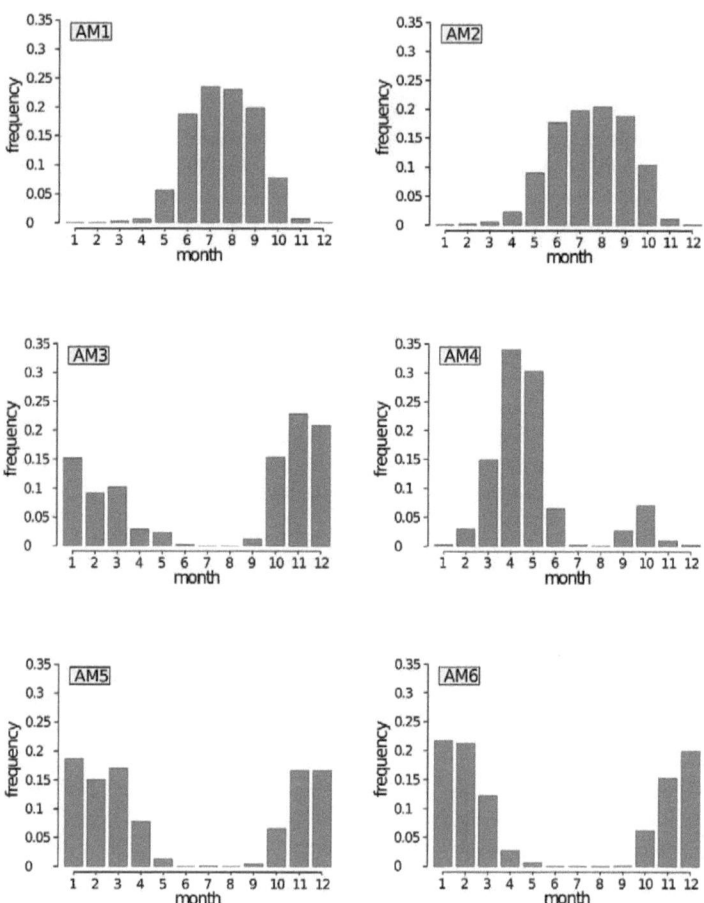

Figure 4.9: Frequencies of the six SOM classes by month.

4.4 Case study: Rhineland region

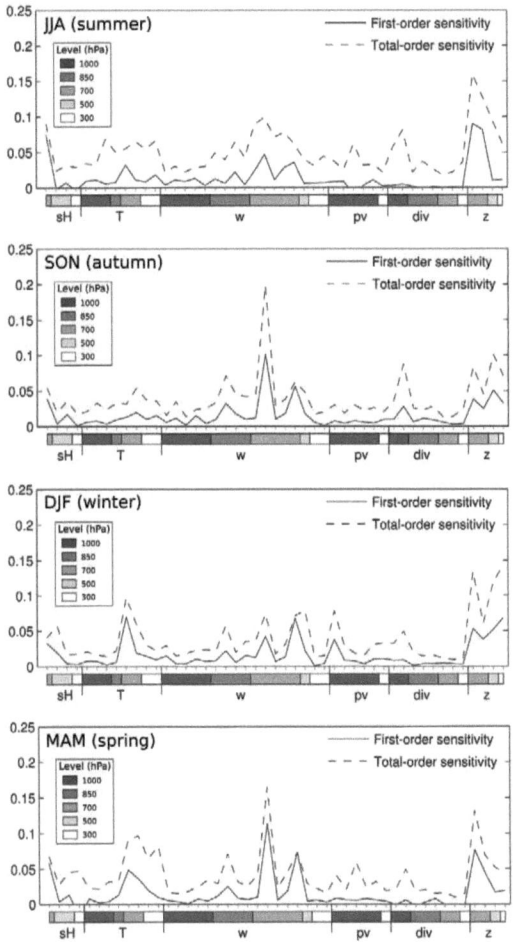

Figure 4.10: Plot of the total- and first-order sensitivity derived from the trained ANN. Every individual lattice of the optimized predictor set (temperature T, specific humidity sH, vertical velocity w, potential vorticity pv, divergence div, geopotential height z) is depicted on the abscissa. The corresponding geopotential levels are represented by the horizontal colorbar at the bottom.

outcomes. Most likely the two subsamples will come from different unknown probability density functions of the predictors. The cumulative distributions are compared applying a two-sided Kolmogorov-Smirnov test. In Figure 4.11 the differences between the strong and the weak class of the average Kolmogorov-Smirnov test-statistics over all stations are shown for all predictors. The larger the value the more important the predictor in producing strong precipitation. In general, high influential predictors (total-order indices) also trigger stronger events. However, apparently less important parameters, such as potential vorticity, *pv*, and divergence, *div*, can also play a decisive role in the tails of the prediction.

4.5 Conclusions and Discussion

A new predictor optimization algorithm for precipitation downscaling based on a conditional air mass classification is presented. These air masses were optimised to have a wide range of precipitation distributions. The principle aims of this work are to study the spatial extent of the informative predictor domain and to estimate the corresponding sources of uncertainty. A particular emphasis was placed on the insights of nonlinear predictor interactions with the intent to indicate the sensitivities of the model to the predictors, which information may used to develop more parsimonious models with only small decreases in accuracy.

As shown in Section 4.4.1, allowing for nonlinearities in the screening process exposes clearly defined and physically interpretable predictor domains, which are by far rectangular nor symmetrical. The comparison of the downscaling results, of artificial neural networks (ANN) trained on the optimized predictor domains with those of symmetric ones confirms the improvements achieved by the optimization. Due to the nonlinear screening, the explained variance of a simple ANN downscaling approach could be improved by up to 9%. This value exceeds the gain of explained variance of about 5% reported by Crawford et al. (2007) using a site-specific choice of surrounding grid-boxes. The subsequent global sensitivity analysis (GSA) provided the necessary understanding of the model's sensitivity pattern. Since, about one-fourth of the ANN model output variance is due to interaction effects this should not, if possible, be neglected in the screening process. In this context linear screening methods are insufficient as they neither account for interactions nor for non-additivity as given by many nonlinear prediction algorithms. This also holds true for the widely used sigma-normalized derivative sensitivity analysis, which might lead to wrong conclusions because of the missing interactions effects.

Based on the results from the domain optimization and the subsequent GSA the following conclusions can be drawn for Central Europe:

- Geopotential height (z) was found to be an influencing predictor for all seasons. Dur-

4.5 Conclusions and Discussion

ing autumn and winter the high-tropospheric region (300 hPa) seems to be more important, while there is a pronounced shift downwards to the 500 hPa level in summer and spring. Interestingly, the horizontal extent of the domain is relatively small. Outside of this domain pressure will also correlate with precipitation in the Rhineland; this illustrates that strong correlations does not always equate good prediction.

- Vertical velocity (w) on the 500 hPa level explains almost 30% of the model output variance (including interactions, see Figure 4.10) in the transitional seasons. Together with the contribution on the 850 and 500 hPa level, w seems to be a key element for downscaling precipitation. Unlike other predictors, its domain is sloping upwind across the entire troposphere covering a great part of central europe (see Figure 4.5d).

- Temperature (T) fundamentally differs from other predictors by its dipole domain (see Figure 4.4a) and its distance from the investigation area with the main center of action over Spain. Nevertheless, the mid- to high-tropospheric T values contribute between 10-18% (including interactions, see Figure 4.10) to the total model output variance.

- The role of humidity (sH) appears to be more relevant for summer precipitation when sub-scale processes are dominant. This is also the most likely explanation for the modest horizontal expansion of the domain. The influence of sH shows a pronounced vertical differentiation decreasing from the 700 hPa (8-18%) to the 300 hPa (2-3%) level (see Figure 4.10).

- Potential vorticity (pv) and divergence (div) may be classified as rather insignificant by themselves in particular in summer and spring. During that time of the year these predictors contribute almost entirely only due to interactions effects (see Figure 4.10). For parismonious models these predictors may be removed with little loss of accuracy.

In general, the decompositon of the complete model output variance by a GSA offers a comfortable way to estimate the relative importance of predictors.

Although the downscaling results are convincing, better results may be obtained by additionally including appropriate time lags and scales. Altogether, predictor optimizion by means of conditional air mass classification provides a good way to improve downscaling results. However, the proposed algorithm is rather time consuming, which is caused by the simulated annealing algorithm on the one hand side and the self-organising maps clustering algorithm on the other side. One way to speed up the optimization can be achieved by using alternative clustering algorithm (e.g. k-means). The predictor domains turned out so be mostly compact and relatively smooth. The most difficult domain is for temperature, which shows a dipole. Therefore, a more greedy and efficient optimisation algorithm may be sufficient to find the optimum domains.

4 Natural three-dimensional predictor domains for statistical precipitation downscaling

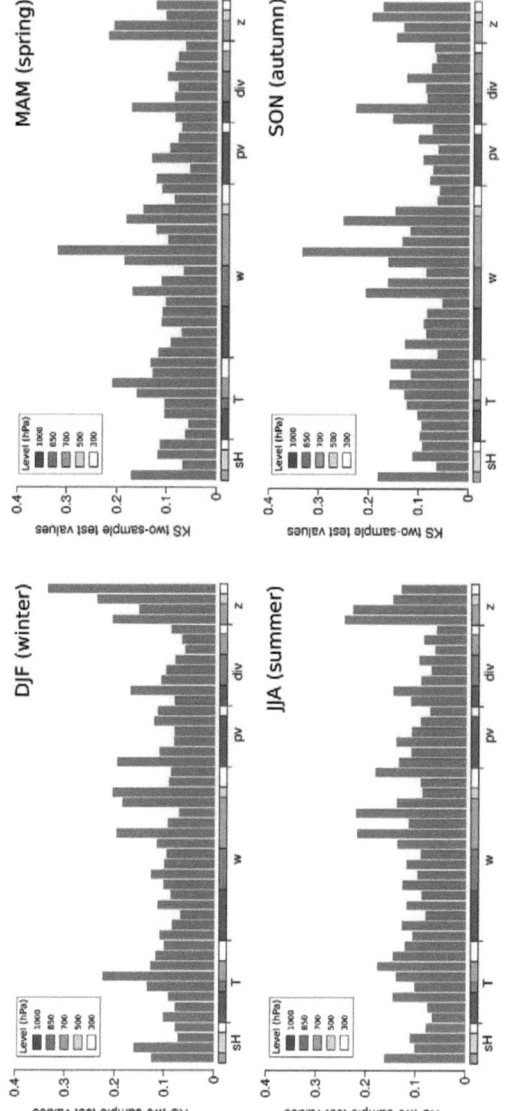

Figure 4.11: The difference between two-sample Kolmogorov-Smirnov test-statistics of the upper 5% tail of the output realization and the remaining 95% for all predictors. The corresponding geopotential levels is represented by the horizontal colorbar at the bottom.

4.5 Conclusions and Discussion

Table 4.3: Explained variance (R^2), RMSE, skill, differences of mean precipitation DP (%), ratio of the average number of dry spells DS (more than 3 days) and the ratio of the average length of dry spells calculated for the validation period for all seasons. Shown are the results for both models M1 (optimized predictors with SOM) and M2 (3x3 grid).

	R^2		RMSE		Skill		DP		No. of DS		Length of DS	
	M1	M2	M1	M2	M1	M2	M1	M2	M1	M2	M1	M2
DJF	0.56	0.53	2.50	2.59	0.77	0.77	-0.07	-0.10	1.01	0.91	0.81	0.84
MAM	0.39	0.39	3.06	2.87	0.67	0.63	0.03	-0.06	0.69	0.74	0.74	0.74
JJA	0.42	0.33	3.79	3.71	0.68	0.57	-0.02	-0.11	0.73	0.82	0.60	0.62
SON	0.50	0.43	3.39	3.56	0.67	0.62	-0.07	-0.03	0.70	0.79	0.78	0.75

Chapter 5

Discussion and conclusions

In the course of this thesis, fundamental aspects in the development and analysis of data-driven prediction algorithms have been discussed. It has been shown that under certain premises, data related models may indeed provide outstanding results, even for complex time series. Depending on the model type and the complexity of signals, different conditions must be met in order to guarantee reliable predictions. Such essentials conditions have been worked out with particular emphasis on important modelling issues like predictor selection, verification and estimation of uncertainty.

5.1 General modelling issues

Nonparametric prediction models are intended to provide statistical inferences on complex systems. With respect to such models, statistical inferences implies learning to generalize from noisy and often nonstationary data. Temporal sequences violate the statistical independence assumption so that observations follow a certain time order. It is assumed that the autocorrelation of consecutive observations is nonzero, thus current observations may also depend on values (both predictor and predictand) of the recent past (Wilks, 2006). A common way to deal with the temporal dependence in non-parametric models is to include lagged input variables (Kecman, 2001). This approach allows one to make also use of ordinary non-parametric static model approaches.

The findings of Chapter 2 confirm that static feedforward networks with tabbed delay lines are able to reproduce temporal structures, if predictors and their corresponding embedding

5 Discussion and conclusions

spaces[1] are selected carefully. Because of the nonlinear nature of ANNs this is not always straightforward as will be discussed in detail in Section 5.3 (also see Chapter 4).

A direct confrontation with a much simpler MLR approach exposed the superior numerical accuracy of such a model. Similar numerical improvements have been also reported by a large number of comparative studies (e.g. Huang et al., 2004; Supharatid, 2003; Riad et al., 2004; Schoof and Pryor, 2001). Systematic errors and biases could not be found either in the mean value or in the extremes. The prediction of extremes require extrapolations of values at unusually large or small levels which turns out to be difficult for nonlinear processes. In comparison with the linear MLR approach, prediction errors do not exponentially increase with the extreme values. This may be due to asymptotic limits of the neuron activation function on the one hand side and the proper rescaling of variables on the other. So it is essential to rescale the inputs so that their variability reflects their importance. Rescaling the predictors and predictand to the same range (e.g. between -1 and +1) together with a sigmoid activation function revealed good results in all case studies.

In discrete time-series modelling current calculations often depend on earlier processing stages. This is difficult to realize with static networks. A possible dynamical model approach on the basis of recurrent NARX was presented in Chapter 3. Obviously, there may be different reasons why the network systematically underestimates snowcover depth. Stochastic components and fast decaying correlations might be two of the main reasons for these systematical errors. Error propagation due to the memory-effect of previous values finally adds up these errors. Despite these undesired error sources, prediction does not diverge and is conditionally stable. As discussed in Section 3.4.1, NARX often tend to produce instabilities, oscillations, bifurcation and chaos due to the high sensitivity to changes of neural weights (Cao and Wang, 2004). More stability can be attained by using a serial-parallel architecture during training as was shown in Chapter 3.

In summary, it can be stated that both models provide better results than comparable linear approaches. According to the formulated measures of model utility (see Chapter 1) we have shown that besides the general ease of model development (low-cost) accurate prediction results are obtained. Even though dynamical networks might more easily lead to unstable solution, they should be prefered for predicting time dependent sequences. This is basically related to the fact, that the propagation of previous values is part of the network structure itself. Optimizing the net weights with the Levenberg-Marquart algorithm leads to a fast and stable convergence. Van der Smagt and Hirzinger (1996) and others showed that ill-conditioning[2] which is a common cause of slow and inaccurate results is not a serious concern to this algorithm. Internal parameters like the number of hidden units and nodes are significantly involved in the prediction accuracy. The optimal number mainly depends

---

[1] In comparison to autoregressive models embedding spaces include all lagged input variables and do not refer only to lagged target values
[2] corresponding to a high condition number

on the complexity of the signal and the amount of noise. As stated by Freeman and Skapura (1991), there are no upper limits if early stopping is applied to prevent overfitting. However, we found that one hidden layer consisting of only a few nodes, not exceeding more than 10 to 25% of the total number of predictors, provides best generalization results. In order to guarantee reliable extrapolations for out-of-sample values and cross validation, a sufficient number of training samples is indispensable. Since the data sets used contained more than 100 times the number of predictors, it was unlikely to suffer from overfitting.

Special attention must be paid to the application of time-delay models if time series are non-stationary whether in terms of seasonality or long term trends. Such a smooth variation requires an extrapolation of the lagged values. This is in general unreliable. Removing these unwanted trends by fitting a polynom is a common way handling also nonlinear deterministic trend (Wilks, 2006). The result suggest that smooth long term trends of climatological and hydrological time series can be ignored if the time series are shorter than ten years (see Chapter 2 and 3).

Owing to the multitude of error sources, resulting times series must be verified in depth. Besides a statistical evaluation, this should include inter-model comparison (see Chapter 4) as well as rigorous assessment of non-parametric approaches versus traditional linear prediction algorithms (see Chapter 2). Usually, time invariant error measures, such as MSE, RMSE and R^2, are employed in order to evaluate the model quality. Despite the importance of spatio-temporal structures of time series the assessments of these characteristics are often neglected. The need of such measures was extensively discussed in Chapter 3 and Chapter 4. Sometimes high correlations have been observed, even though there was a lack in capturing the spatial and temporal structures. Hence, from a modelling point of view it is suggested to combine different time variant and invariant measures.

5.2 Nonlinear determinism

In case prediction results show a systematical error as in Section 3.5, it is necessary to ask if these fluctuations reveal nontrivial structures in the serial correlations. While a priori fluctuations seem very likely to origin from a stochastic Gaussian process, there is no evidence for possible nonlinear structures in the dynamics. Surrogate data testing as proposed by Schreiber and Schmitz (1996) has become a popular tool to address such a question. Our results suggested that residuals are a linear stochastic process so that it can be assumed that the most part of the nonlinear determinism was captured by the NARX model. The fact that no nonlinear components were found does not necessarily prove the absence of such signals. To our best knowledge, the underestimation is mainly attributable to external stochastic processes. In order to confirm this assumption, more constrained surrogates (e.g. by simulated annealing) are advisable. Even though surrogates testing was not ap-

5 Discussion and conclusions

plied earlier to scrutinize model residuals, it seems to provide useful information on model quality.

5.3 Predictor optimization

The scientific community has made great efforts to develop and optimize non-parametric prediction algorithms. But only a few studies are dedicated to predictor screening processes, e.g. Wilby and Wigley (2000); Cavazos (2005). Although predictors are intended to be optimized for nonlinear prediction algorithms, most approaches are entirely based on linear assumptions (see Chapter 2), such as linear correlation (Wilby and Wigley, 2000; Kwak and Choi, 2002) or linear transformation (Hyvärinen and Oja, 2000). Obviously, this is accompanied by a loss of important information which might be useful for successful prediction. In Chapter 4 a new optimization algorithm for high dimensional potential predictor domains was proposed allowing for nonlinearities in the screening process. Due to the nonlinear screening, the predictive power of a simple ANN downscaling approach could be improved by up to 9% compared to a linear optimization. Most surprisingly, the smooth shapes of the optimized predictor domains can be well related to mesoscale atmospheric processes. The results support the hypothesis that regular predictor domains may not be optimal for establishing linkages between atmospheric characteristics and local scale phenomena. Even the horizontal extent of the domains are relatively small despite slow decaying correlation lengths. Thus strong correlation does not always imply good prediction. Overall, the result illustrate that neglecting nonlinear predictor interactions in the screening process significantly reduces important information and hence the quality of the prediction. It is claimed that linear screening methods are insufficient as they neither account for interactions nor for nonadditivity as given by many nonlinear prediction algorithms.

While the application was restricted to just one case study, it is planned to validate the algorithm further on in the near future. In order to cover different climatological settings, the methods will be used to optimize predictors for precipitation downscaling in Oman and Svalbard (Arctic).

5.4 Global sensitivity analysis

Since data-driven methods distribute information in the model structure itself, direct assignment of processes to specific model parameters turn out to be difficult. In recent years some attempts have been made to derive symbolic rules from model parameters (Taha and Ghosh, 1999; Jain et al., 2004; Wilby et al., 2003; Engelbrecht and Cloete, 1999). However, these approaches do not allow causal inferences to assess the significance of predictors

5.4 Global sensitivity analysis

(model inputs). The estimation of first- and higher-order effects with a GSA is suggested in connection with data-driven methods. This permits inferences on different sources of uncertainty in the model input by decomposing the variance of the model output. The quantification of these effects help to understand the model's complete sensitivity pattern. Based on this knowledge irrelevant predictors can be pruned, thus effectively reducing the number of predictors for more parsimonious models.

The efficiency of the GSA to non-parametric models was first demonstrated by means of the hydrological time series, where processes are well known a priori (Chapter 2). At least qualitatively, outcomes are in good agreement with the prevailing observed processes. About 96.2% of the output variance can be explained by individual predictors, whereas the autoregressive fraction (antedecent water levels) takes in about 49.2%. The high proportion is the consequence of the slowly decaying autocorrelation function. The weak non-additive character brings forth the idea of further reducing model complexity by using a simple additive model. Interaction effects are more important in the ANN precipitation downscaling model (Chapter 4). Here, interactions have an impact of up to 27% on the output variablility. Some predictors only contribute to the output variance in combination with other factors. Many linear screening methods might declare such predictors as non-significant. This stresses again the need of appropriate nonlinear predictor optimization for such models (Section 5.3). These results indicate, that with increasing time series complexity, predictor interactions become more and more important and cannot neglected.

Appendix A

Fourier based surrogates

To test the null hypothesis of an underlying stationary linear stochastic Gaussian process a set of constrained realisations are created. Lets denote the measured time series as s_n. The iterative amplitude adapted Fourier transform starts with a random shuffle of the data. In a first step the power spectrum $|S_k|^2$ of the original time series is calculated

$$|S_k|^2 = \left| \frac{1}{\sqrt{N}} \sum_{n=0}^{N-1} s_n e^{i 2\pi k n / N} \right|^2 \tag{A.1}$$

Assuming that the linear properties of the measured time series are specified by the power spectrum, the Fourier transform of the initial random time series is calculated and replaced by those of the original time series $|S_k|^2$. The phases $e^{i\phi_k}$ remain unaltered, where the ϕ_k are uniformly distributed in $[0, 2\pi[$. The new time series is transformed back to the time domain having the same power spectrum but a different distribution.

$$\bar{s}_n = \frac{1}{\sqrt{N}} \sum_{k=0}^{N-1} e^{i\phi_k} |S_k| e^{-i 2\pi k n / N} \tag{A.2}$$

The surrogate time series $\bar{s}_n$ now contains random numbers with the prescribed power spectrum. The sequence is then rescaled to the empirical distribution by a simple rank ordering. Substituting the ranked values of the sequence $\bar{s}_n$ by the same ranked values of the reference sequence s_n gives the rescaled sequence. The adjustment of the distribution will alter the power spectrum, so that both steps must be repeated until a convergence threshold is reached.

Appendix B

Locally constant predictor in phase space

The locally constant predictor is a simple prediction algorithm which exploits deterministic structures in the signal. Pure dynamical systems are usually described by discrete[1] time maps,

$$x_{n+1} = F(x_n). \tag{B.1}$$

In reality one does not measure the actual states x_n, but rather has scalar measurements,

$$s_n = s(x_n) \quad n = 1, \cdots, N, \tag{B.2}$$

in which the measurement function s and F are unknown. The sequence is transformed by a delay reconstruction to obtain an equivalent vector in phase space,

$$\mathbf{s}_n = (s_{n-(m-1)\tau}, s_{n-(m-2)\tau}, \cdots, s_{n-\tau}, s_n), \tag{B.3}$$

with the delay time τ and the embedding dimension m. In order to predict the future of a point $s_{N+\Delta n}$ in delay embedding space, the algorithm searches for the closest elements in the neighbourhood $\odot_\varepsilon(\mathbf{s}_N)$ of radius ε around the point $\mathbf{s}_n$. Finally, the prediction is obtained by taking the average over all these points $\mathbf{s}_n \in \odot_\varepsilon(\mathbf{s}_N)$,

---
[1] or first-order ordinary differenctial equations

B Locally constant predictor in phase space

$$\hat{s}_{N+\Delta n} = \frac{1}{|\odot_\varepsilon(\mathbf{s}_N)|} \sum_{\mathbf{s}_n \in \odot_\varepsilon(\mathbf{s}_N)} s_{n+\Delta n}. \qquad (B.4)$$

The expression $|\odot_\varepsilon(\mathbf{s}_N)|$ denotes the number of points within the defined neigbourhood $\odot_\varepsilon(\mathbf{s}_N)$. If no neighbours are found, the ε will be increased until at least one point is found.

Appendix C

Snow-cover maps

C Snow-cover maps

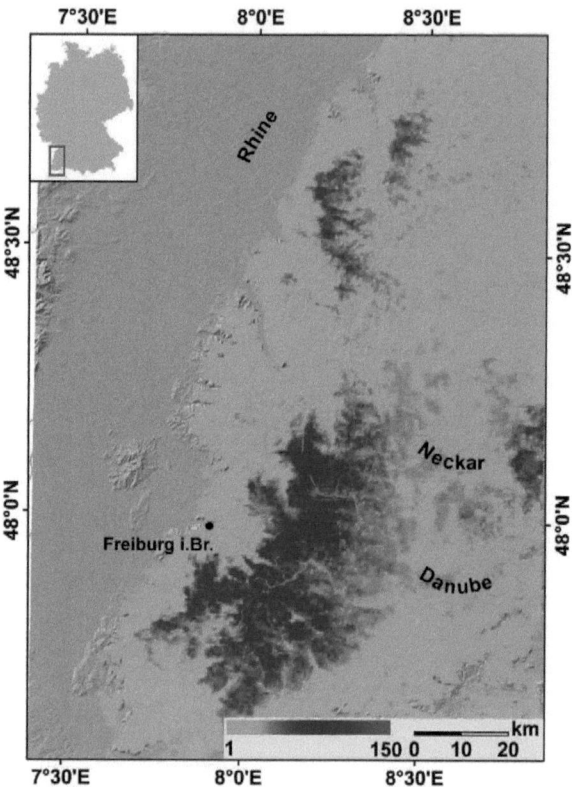

Figure C.1: The observed seasonal maps of snow cover days for the validation periods (1994-2003) are displayed. Only heights above 250 m are considered. The inset shows the location of the study area within Germany.

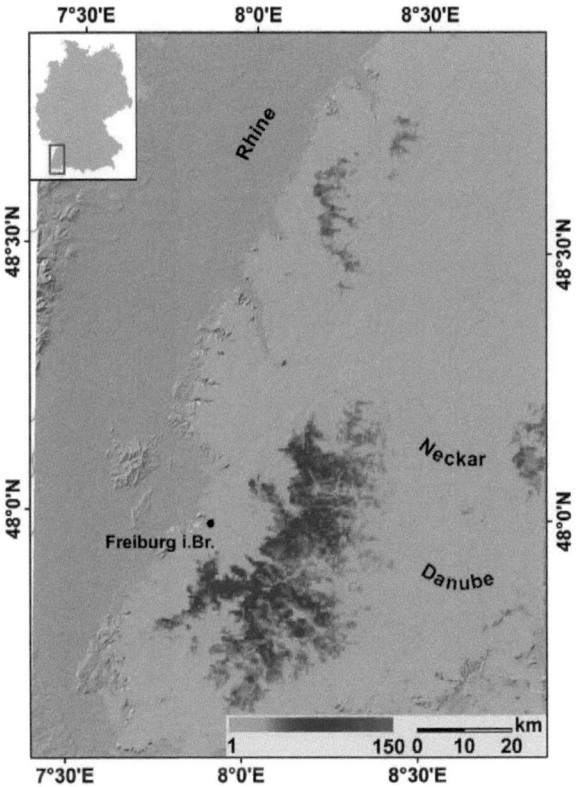

Figure C.2: The modelled seasonal maps of snow cover days for the validation periods (1994-2003) are displayed. Only heights above 250 m are considered. The inset shows the location of the study area within Germany.

C Snow-cover maps

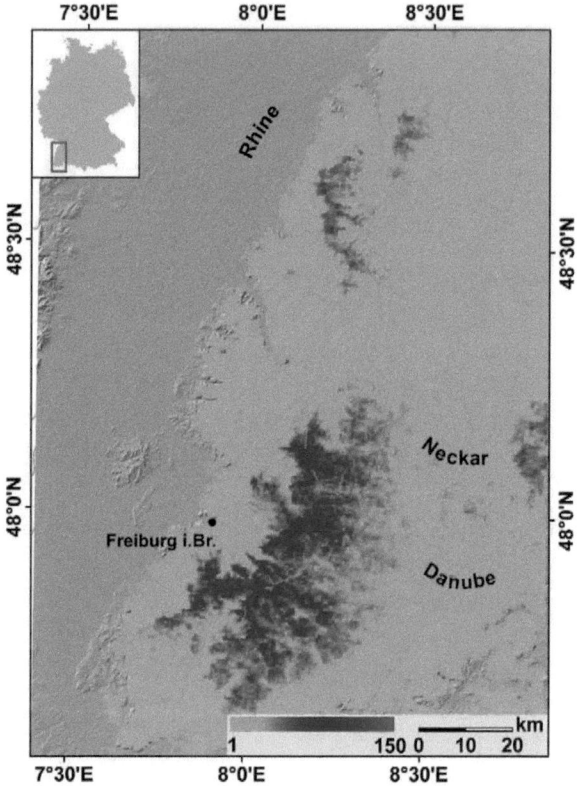

Figure C.3: Snow-cover maps for the MF scenario for the decade 2021-2030; (for a detailed description of the scenario data, see Section 3.3). Only heights above 250 m are considered. The inset shows the location of the study area within Germany.

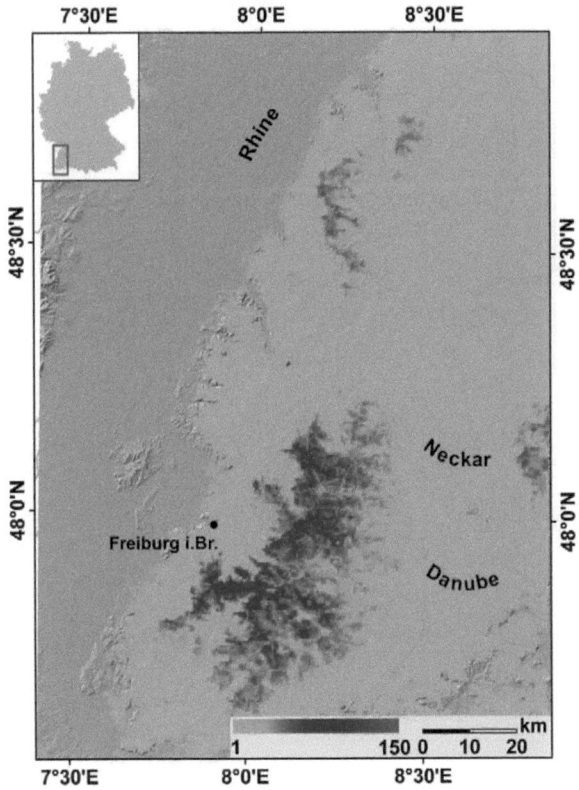

Figure C.4: Snow-cover maps for the MT scenario for the decade 2021-2030; (for a detailed description of the scenario data, see Section 3.3). Only heights above 250 m are considered. The inset shows the location of the study area within Germany.

C Snow-cover maps

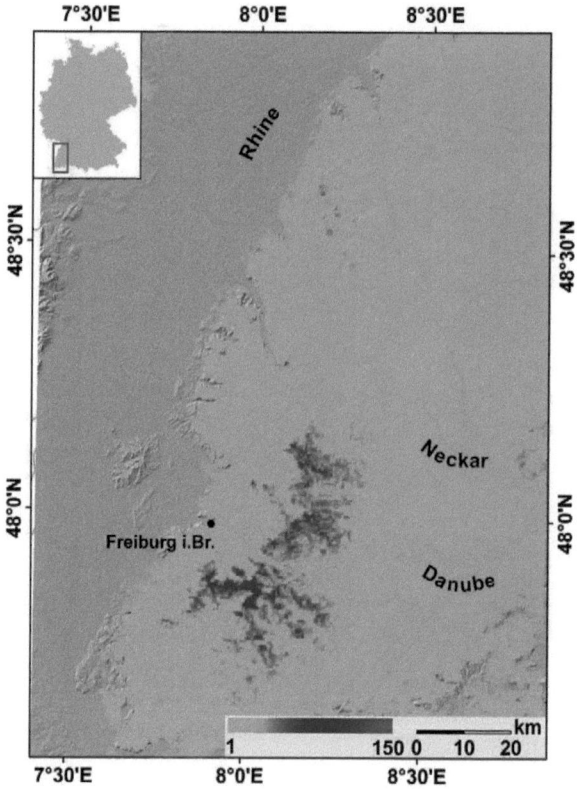

Figure C.5: Snow-cover maps for the MF scenario for the decade 2041-2050; (for a detailed description of the scenario data, see Section 3.3). Only heights above 250 m are considered. The inset shows the location of the study area within Germany.

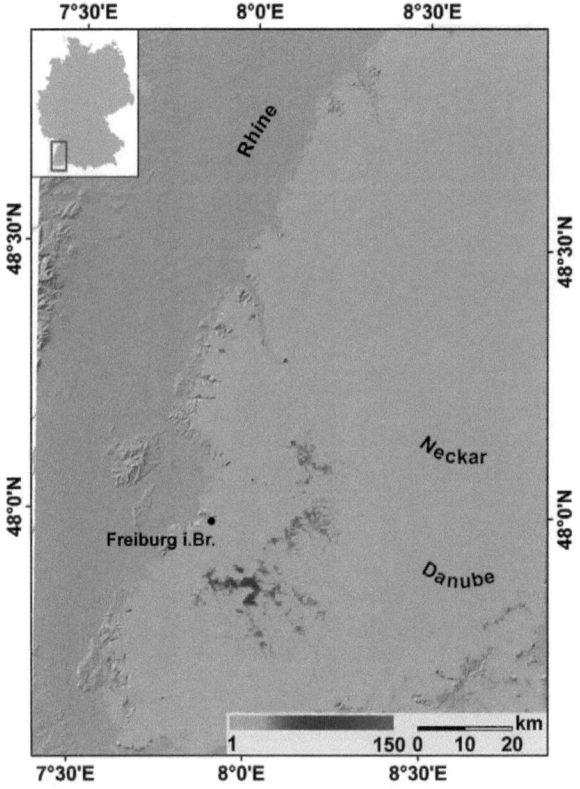

Figure C.6: Snow-cover maps for the MT scenario for the decade 2041-2050; (for a detailed description of the scenario data, see Section 3.3). Only heights above 250 m are considered. The inset shows the location of the study area within Germany.

Appendix D

Estimated monthly changes in the number of snow-cover days

D Estimated monthly changes in the number of snow-cover days

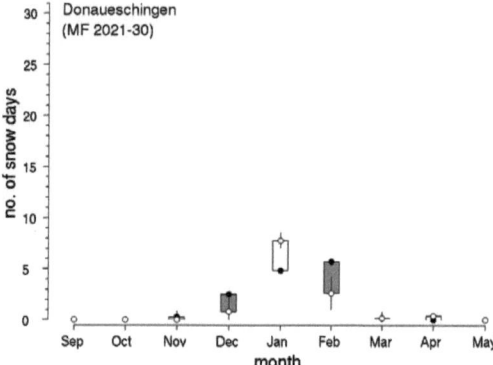

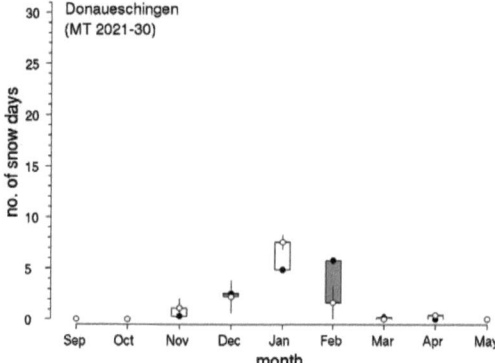

Figure D.1: Monthly snow-cover days' changes at the location Donaueschingen by 2021-30 for different scenarios (MF = wet, MT = dry; for a detailed description of the scenario data, see Section 3.3. The plot is described and discussed in Section 3.5.

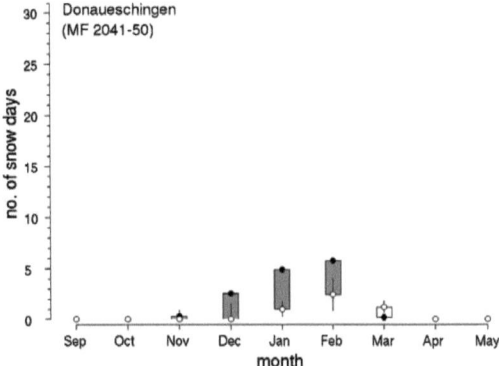

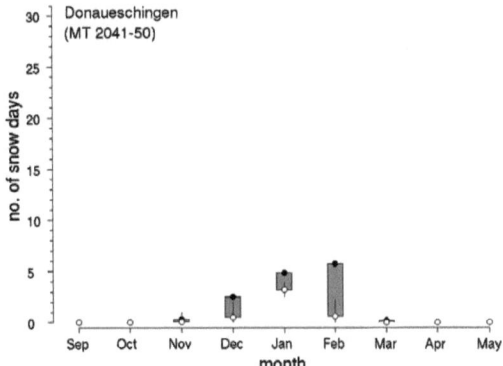

Figure D.2: Monthly snow-cover days' changes at the location Donaueschingen by 2041-50 for different scenarios (MF = wet, MT = dry; for a detailed description of the scenario data, see Section 3.3. The plot is described and discussed in Section 3.5.

D Estimated monthly changes in the number of snow-cover days

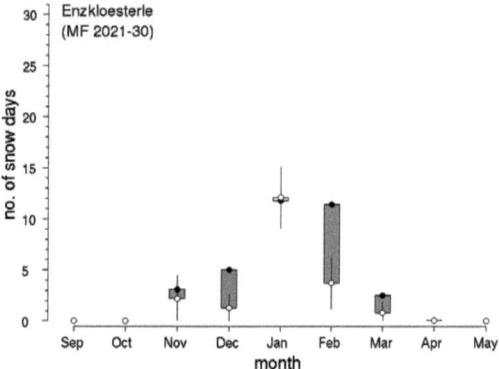

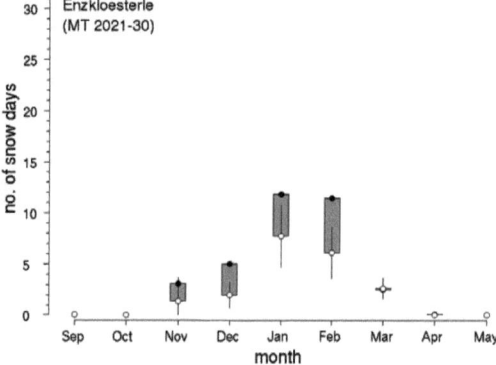

Figure D.3: Monthly snow-cover days' changes at the location Enzkloesterle by 2021-30 for different scenarios (MF = wet, MT = dry; for a detailed description of the scenario data, see Section 3.3. The plot is described and discussed in Section 3.5.

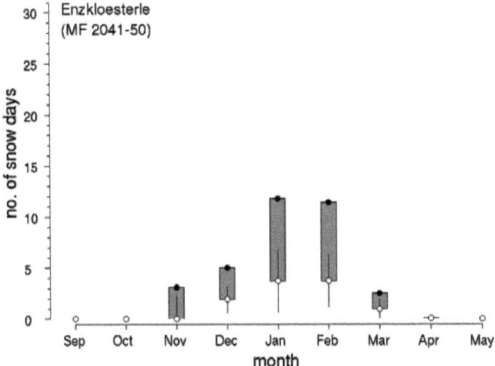

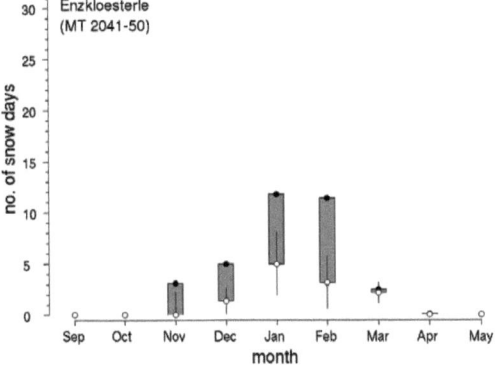

Figure D.4: Monthly snow-cover days' changes at the location Enzkloesterle by 2041-50 for different scenarios (MF = wet, MT = dry; for a detailed description of the scenario data, see Section 3.3. The plot is described and discussed in Section 3.5.

D Estimated monthly changes in the number of snow-cover days

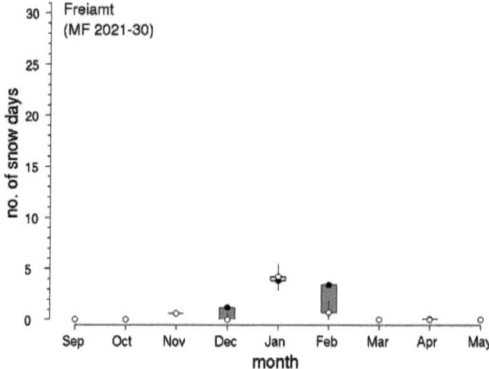

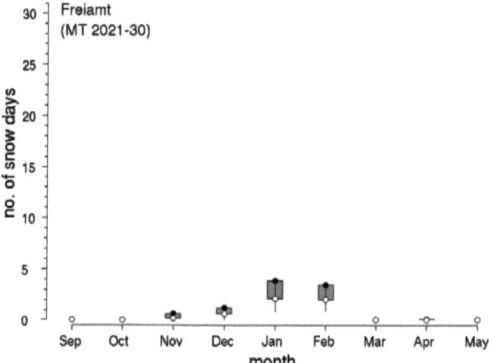

Figure D.5: Monthly snow-cover days' changes at the location Freiamt by 2021-30 for different scenarios (MF = wet, MT = dry; for a detailed description of the scenario data, see Section 3.3. The plot is described and discussed in Section 3.5.

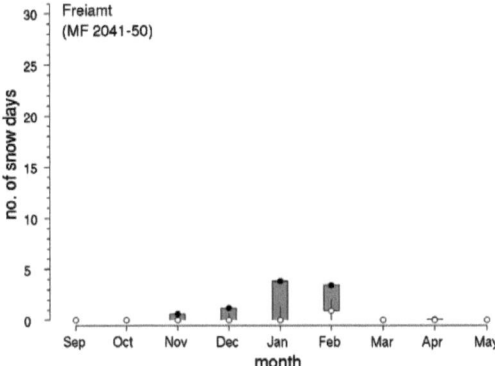

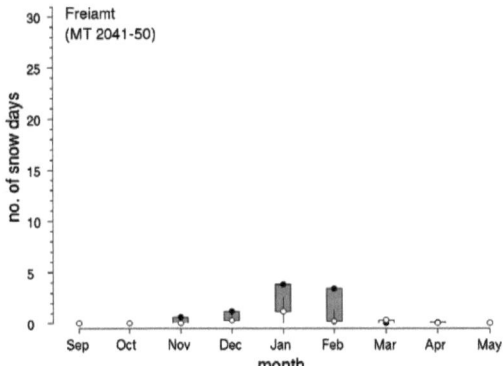

Figure D.6: Monthly snow-cover days' changes at the location Freiamt by 2041-50 for different scenarios (MF = wet, MT = dry; for a detailed description of the scenario data, see Section 3.3. The plot is described and discussed in Section 3.5.

D Estimated monthly changes in the number of snow-cover days

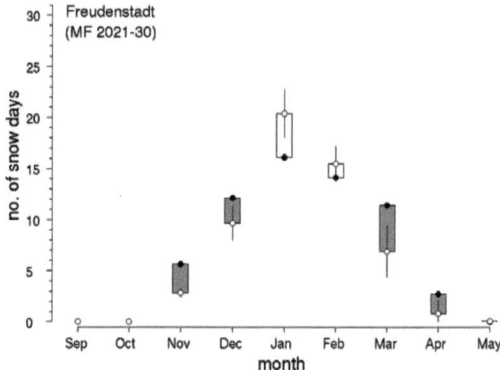

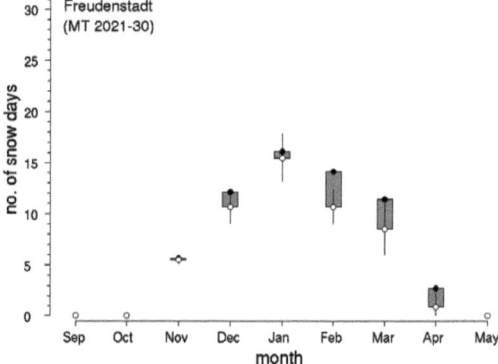

Figure D.7: Monthly snow-cover days' changes at the location Freudenstadt by 2021-30 for different scenarios (MF = wet, MT = dry; for a detailed description of the scenario data, see Section 3.3. The plot is described and discussed in Section 3.5.

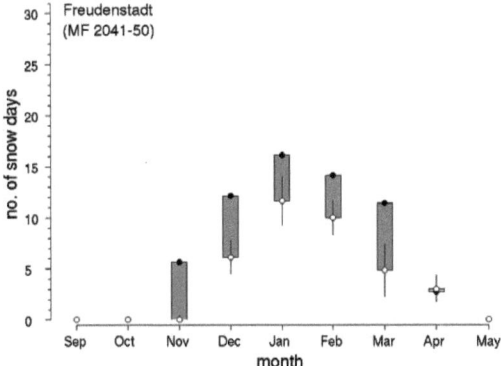

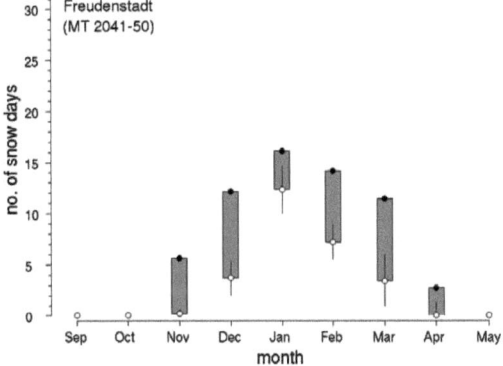

Figure D.8: Monthly snow-cover days' changes at the location Freudenstadt by 2041-50 for different scenarios (MF = wet, MT = dry; for a detailed description of the scenario data, see Section 3.3. The plot is described and discussed in Section 3.5.

D Estimated monthly changes in the number of snow-cover days

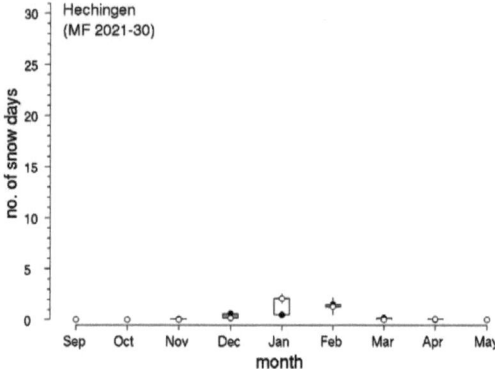

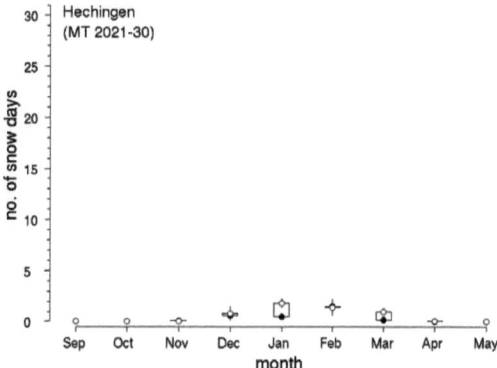

Figure D.9: Monthly snow-cover days' changes at the location Hechingen by 2021-30 for different scenarios (MF = wet, MT = dry; for a detailed description of the scenario data, see Section 3.3. The plot is described and discussed in Section 3.5.

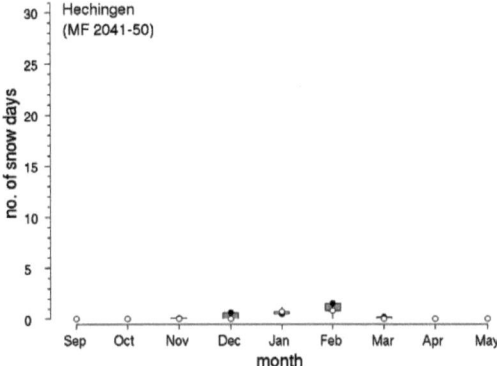

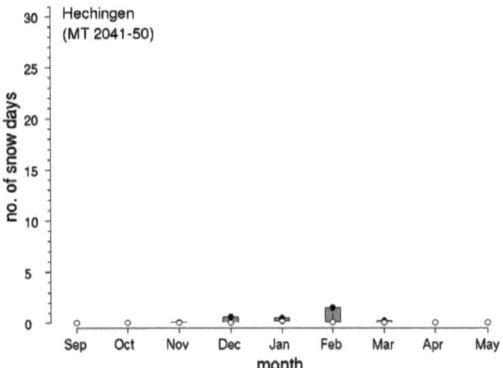

Figure D.10: Monthly snow-cover days' changes at the location Hechingen by 2041-50 for different scenarios (MF = wet, MT = dry; for a detailed description of the scenario data, see Section 3.3. The plot is described and discussed in Section 3.5.

D Estimated monthly changes in the number of snow-cover days

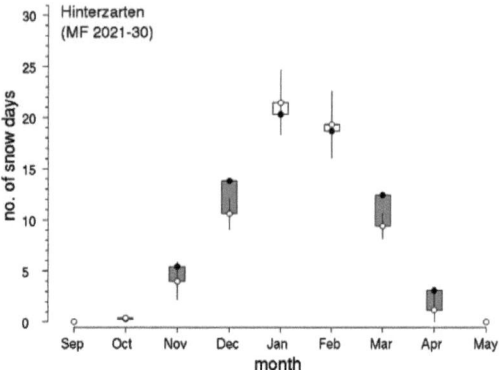

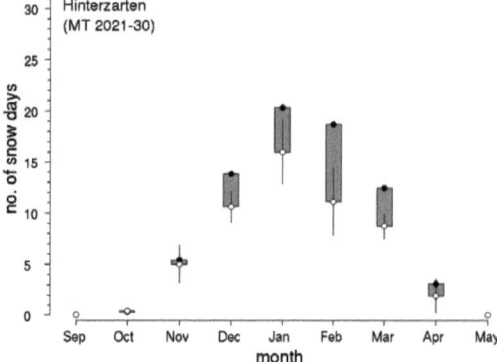

Figure D.11: Monthly snow-cover days' changes at the location Hinterzarten by 2021-30 for different scenarios (MF = wet, MT = dry; for a detailed description of the scenario data, see Section 3.3. The plot is described and discussed in Section 3.5.

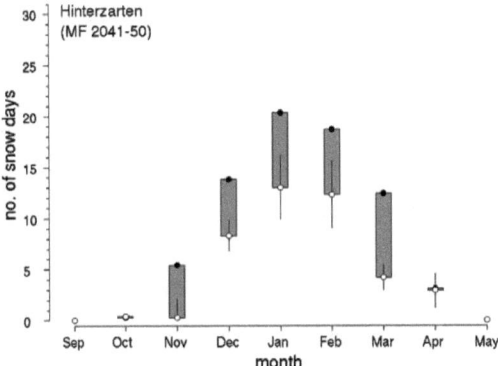

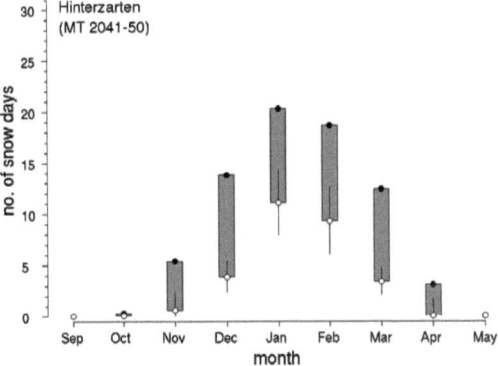

Figure D.12: Monthly snow-cover days' changes at the location Hinterzarten by 2041-50 for different scenarios (MF = wet, MT = dry; for a detailed description of the scenario data, see Section 3.3. The plot is described and discussed in Section 3.5.

D Estimated monthly changes in the number of snow-cover days

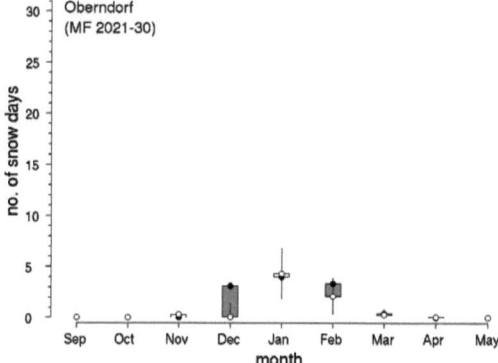

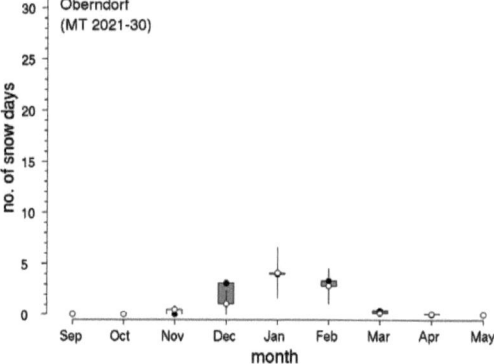

Figure D.13: Monthly snow-cover days' changes at the location Oberndorf by 2021-30 for different scenarios (MF = wet, MT = dry; for a detailed description of the scenario data, see Section 3.3. The plot is described and discussed in Section 3.5.

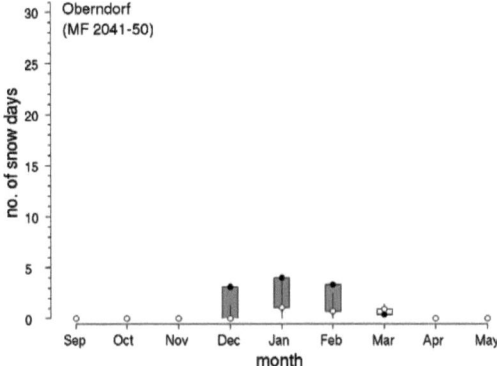

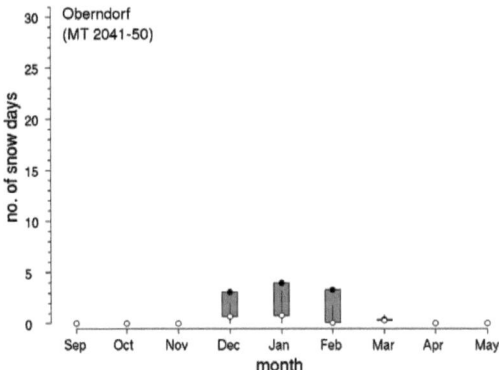

Figure D.14: Monthly snow-cover days' changes at the location Oberndorf by 2041-50 for different scenarios (MF = wet, MT = dry; for a detailed description of the scenario data, see Section 3.3. The plot is described and discussed in Section 3.5.

D Estimated monthly changes in the number of snow-cover days

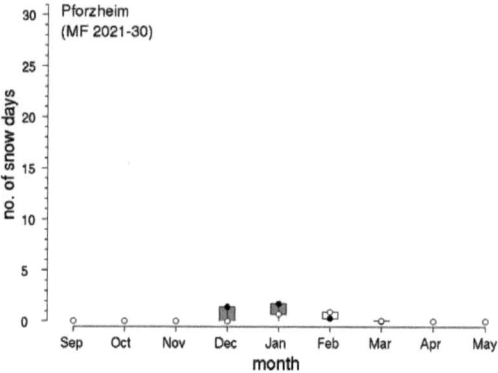

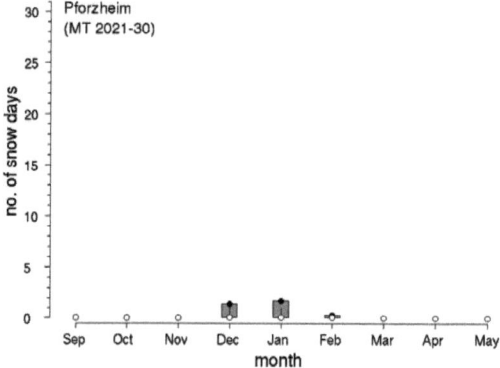

Figure D.15: Monthly snow-cover days' changes at the location Pforzheim by 2021-30 for different scenarios (MF = wet, MT = dry; for a detailed description of the scenario data, see Section 3.3. The plot is described and discussed in Section 3.5.

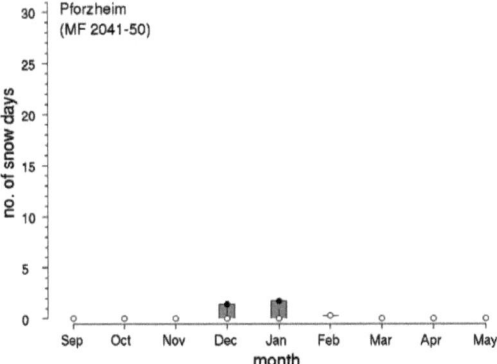

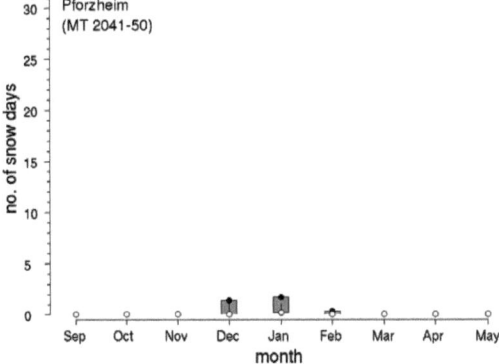

Figure D.16: Monthly snow-cover days' changes at the location Pforzheim by 2041-50 for different scenarios (MF = wet, MT = dry; for a detailed description of the scenario data, see Section 3.3. The plot is described and discussed in Section 3.5.

D Estimated monthly changes in the number of snow-cover days

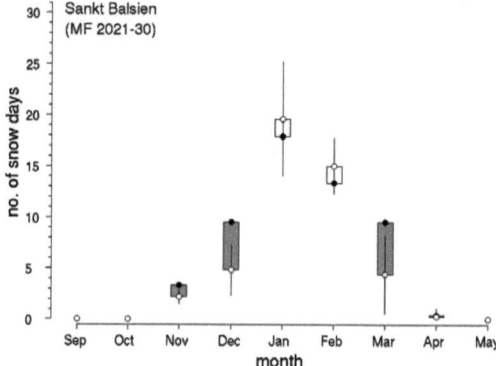

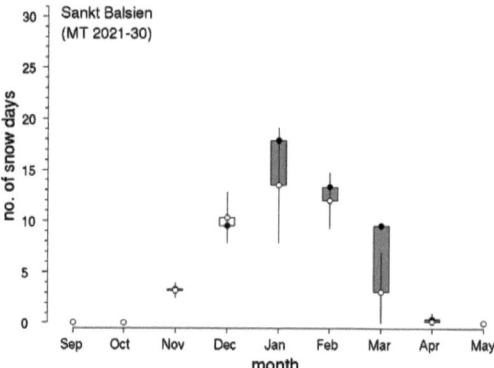

Figure D.17: Monthly snow-cover days' changes at the location Sankt Blasien by 2021-30 for different scenarios (MF = wet, MT = dry; for a detailed description of the scenario data, see Section 3.3. The plot is described and discussed in Section 3.5.

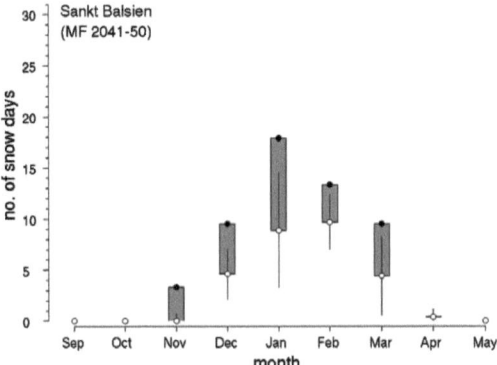

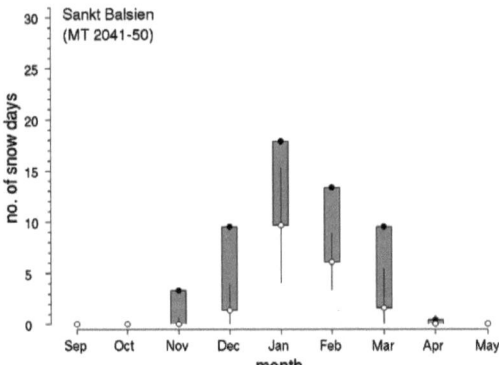

Figure D.18: Monthly snow-cover days' changes at the location Sankt Blasien by 2041-50 for different scenarios (MF = wet, MT = dry; for a detailed description of the scenario data, see Section 3.3. The plot is described and discussed in Section 3.5.

D Estimated monthly changes in the number of snow-cover days

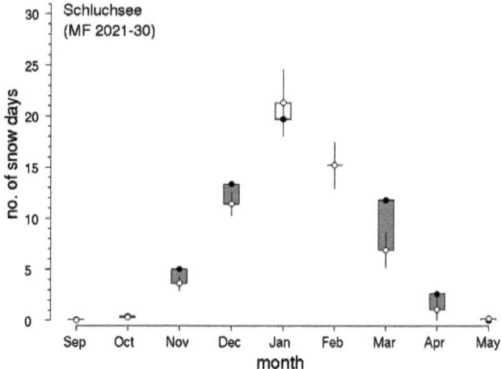

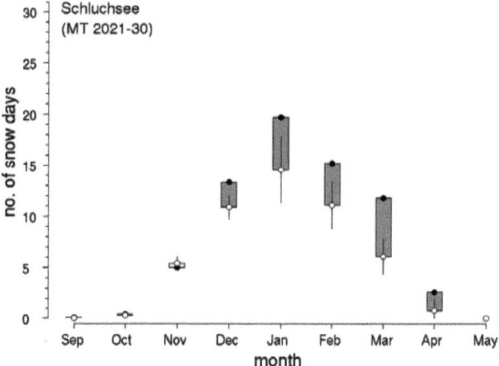

Figure D.19: Monthly snow-cover days' changes at the location Schluchsee by 2021-30 for different scenarios (MF = wet, MT = dry; for a detailed description of the scenario data, see Section 3.3. The plot is described and discussed in Section 3.5.

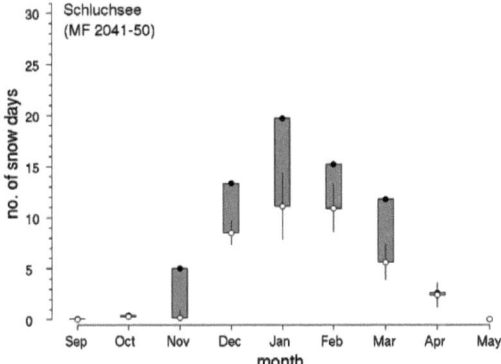

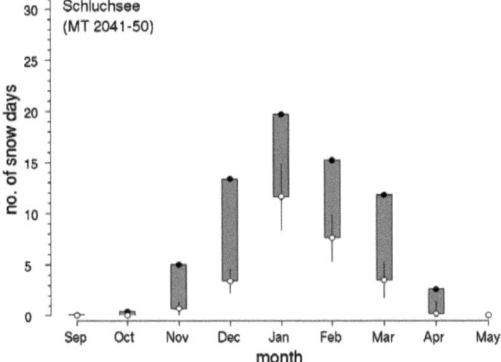

Figure D.20: Monthly snow-cover days' changes at the location Schluchsee by 2041-50 for different scenarios (MF = wet, MT = dry; for a detailed description of the scenario data, see Section 3.3. The plot is described and discussed in Section 3.5.

D Estimated monthly changes in the number of snow-cover days

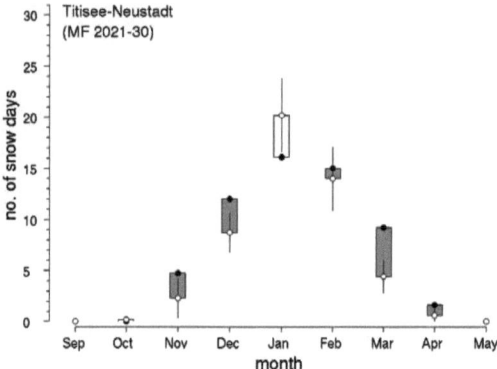

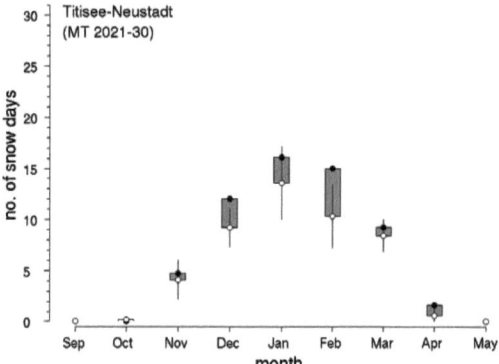

Figure D.21: Monthly snow-cover days' changes at the location Titisee-Neustadt by 2021-30 for different scenarios (MF = wet, MT = dry; for a detailed description of the scenario data, see Section 3.3. The plot is described and discussed in Section 3.5.

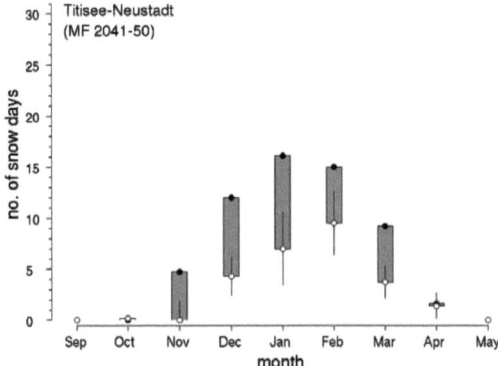

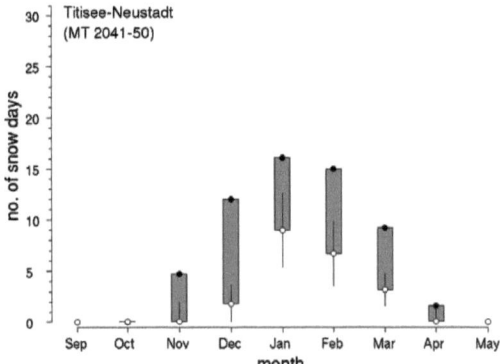

Figure D.22: Monthly snow-cover days' changes at the location Titisee-Neustadt by 2041-50 for different scenarios (MF = wet, MT = dry; for a detailed description of the scenario data, see Section 3.3. The plot is described and discussed in Section 3.5.

D Estimated monthly changes in the number of snow-cover days

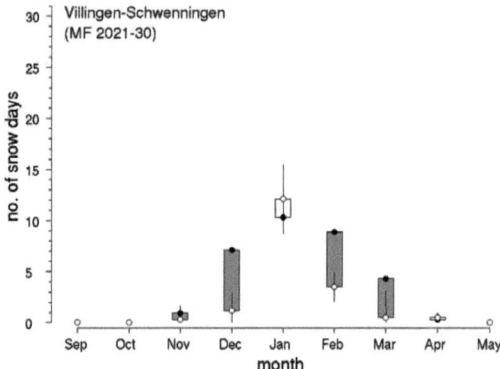

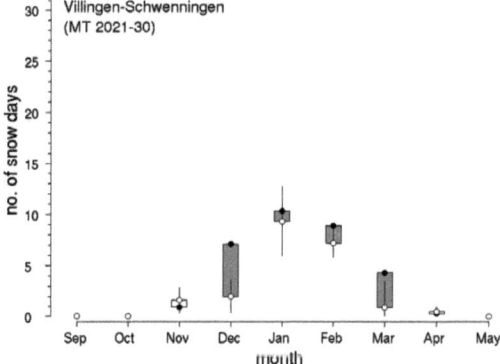

Figure D.23: Monthly snow-cover days' changes at the location Villingen-Schwenningen by 2021-30 for different scenarios (MF = wet, MT = dry; for a detailed description of the scenario data, see Section 3.3. The plot is described and discussed in Section 3.5.

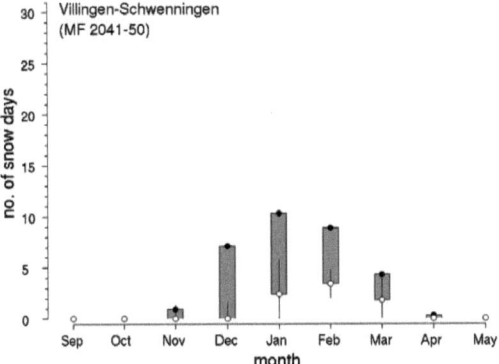

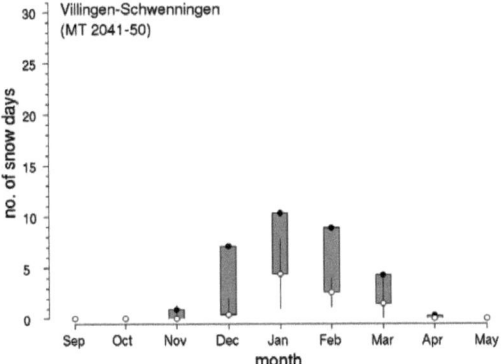

Figure D.24: Monthly snow-cover days' changes at the location Villingen-Schwenningen by 2041-50 for different scenarios (MF = wet, MT = dry; for a detailed description of the scenario data, see Section 3.3. The plot is described and discussed in Section 3.5.

List of Figures

2.1	Schematic illustration of an artificial neuron	13
2.2	Architecture of a multilayer feed-forward network with supervised learning .	15
2.3	AIC and BIC scores of the ANN model	16
2.4	Location of the study area in southernmost Patagonia	19
2.5	Time series of daily mean values and of water level, precipitation, temperature, incoming shortwave radiation and wind speed; daily averages	21
2.6	Statistical analysis of the ANN predictors	22
2.7	Error trajectory of training, test and validation	25
2.8	Correlation coefficient of the MLR model for different time and precipitation time lags	25
2.9	Comparison of actual and predicted water levels of the ANN model for training and validation phases	26
2.10	Comparison of actual and predicted water levels of the MLR model for training and validation phases	26
2.11	Observed and modelled hydrograph of ANN and MLR of a single event . . .	28
2.12	Observed and modelled hydrograph of ANN and MLR of an individual flood event	28
2.13	Observed and modelled hydrograph of ANN and MLR during a period of air temperature below freezing point	28
2.14	Main effects of the meteorological predictors on the runoff	30
2.15	Total effects of the meteorological predictors on the runoff	30
3.1	Digital Elevation Model of the Black Forest mountain region	36
3.2	RMSE between model output and measured snow depth plotted against the delay of the input variables.	42
3.3	Mean fractional snow mask	44
3.4	Nonlinear prediction error	46
3.5	Correlation matrices between weather stations	48
3.6	Transfer function between fractional snow mask and number of snow days .	49

List of Figures

3.7 Observed and modelled seasonal maps of snow cover days for the validation period 50
3.8 Monthly snow-cover days' changes at the location Feldberg by 2021-30 51
3.9 Monthly snow-cover days' changes at the location Feldberg by 2041-50 52
3.10 Seasonal maps of snow cover days for the decade 2021-2030 55
3.11 Seasonal maps of snow cover days for the decade 2021-2030 56
3.12 Seasonal maps of snow cover days for the decade 2041-2050 57
3.13 Seasonal maps of snow cover days for the decade 2041-2050 58

4.1 Flowchart of the proposed predictor optimization process 62
4.2 Digital Elevation Model of the Rheinland region 68
4.3 Cumulative densitiy distribution of precipitation amounts for each of the six airmasses at the weather station Aachen. 69
4.4 Three-dimensional perspectives of the optimized predictor domain 70
4.5 Three-dimensional perspectives of the optimized predictor domain 71
4.6 Mean sea level pressure fields and specific humidity at the 500 hPa level for AM1 and AM2 73
4.7 Mean sea level pressure fields and specific humidity at the 500 hPa level for AM3 and AM4 74
4.8 Mean sea level pressure fields and specific humidity at the 500 hPa level for AM5 and AM6 75
4.9 Frequencies of the six SOM classes by month. 78
4.10 Total- and first-order sensitivity derived from the trained ANN 79
4.11 Kolmogorov-Smirnov test-statistic of the ANN factor mapping 82

C.1 Observed and modelled seasonal maps of snow cover days for the validation period 96
C.2 Observed and modelled seasonal maps of snow cover days for the validation period 97
C.3 Seasonal maps of snow cover days for the decade 2021-2030 98
C.4 Seasonal maps of snow cover days for the decade 2021-2030 99
C.5 Seasonal maps of snow cover days for the decade 2041-2050 100
C.6 Seasonal maps of snow cover days for the decade 2041-2050 101

D.1 Monthly snow-cover days' changes at the location Donaueschingen by 2021-30 104
D.2 Monthly snow-cover days' changes at the location Donaueschingen by 2041-50 105
D.3 Monthly snow-cover days' changes at the location Enzkloesterle by 2021-30 . 106
D.4 Monthly snow-cover days' changes at the location Enzkloesterle by 2041-50 . 107
D.5 Monthly snow-cover days' changes at the location Freiamt by 2021-30 108
D.6 Monthly snow-cover days' changes at the location Freiamt by 2041-50 109
D.7 Monthly snow-cover days' changes at the location Freudenstadt by 2021-30 . 110

List of Figures

D.8 Monthly snow-cover days' changes at the location Freudenstadt by 2041-50 . 111
D.9 Monthly snow-cover days' changes at the location Hechingen by 2021-30 . . . 112
D.10 Monthly snow-cover days' changes at the location Hechingen by 2041-50 . . . 113
D.11 Monthly snow-cover days' changes at the location Hinterzarten by 2021-30 . 114
D.12 Monthly snow-cover days' changes at the location Hinterzarten by 2041-50 . 115
D.13 Monthly snow-cover days' changes at the location Oberndorf by 2021-30 . . . 116
D.14 Monthly snow-cover days' changes at the location Oberndorf by 2041-50 . . . 117
D.15 Monthly snow-cover days' changes at the location Pforzheim by 2021-30 . . . 118
D.16 Monthly snow-cover days' changes at the location Pforzheim by 2041-50 . . . 119
D.17 Monthly snow-cover days' changes at the location Sankt Blasien by 2021-30 . 120
D.18 Monthly snow-cover days' changes at the location Sankt Blasien by 2041-50 . 121
D.19 Monthly snow-cover days' changes at the location Schluchsee by 2021-30 . . . 122
D.20 Monthly snow-cover days' changes at the location Schluchsee by 2041-50 . . . 123
D.21 Monthly snow-cover days' changes at the location Titisee-Neustadt by 2021-30 124
D.22 Monthly snow-cover days' changes at the location Titisee-Neustadt by 2041-50 125
D.23 Monthly snow-cover days' changes at the location Villingen-Schwenningen
by 2021-30 126
D.24 Monthly snow-cover days' changes at the location Villingen-Schwenningen
by 2041-50 127

List of Tables

2.1 Description and time lags of the predictors used in the ANN model 20
2.2 Periods of training, test and validation sets 20
2.3 Model performance indices of the ANN and the MLR models for training and validation 24

3.1 Location and altitude of the Stations 37
3.2 Yearly means of the predictors at the station Feldberg 39
3.3 Yearly means of the predictors at the station Feldberg 45
3.4 Mean changes in snow days for different decades and scenarios 53

4.1 SA paramter set. 66
4.2 Characteristics of air masses for each class. Shown are the mean values of sH and w in the corresponding subdomains. Also given is the mean difference in surface temperature DT and standard deviation (in bracketes) of the T dipole. To gain a detailed understanding also the local mean rain rate RR and fraction of rainy days RD are presented. 76
4.3 Explained variance (R^2), RMSE, skill, differences of mean precipitation DP (%), ratio of the average number of dry spells DS (more than 3 days) and the ratio of the average length of dry spells calculated for the validation period for all seasons. Shown are the results for both models M1 (optimized predictors with SOM) and M2 (3x3 grid). 83

Bibliography

Abrahart, R. J., P. E. Kneale, and L. M. See, 2004: *Neural networks for hydrological modelling*. Taylor & Francis.

Akaike, H., 1974: A new look at the statistical model identification. *IEEE Transactions on Automatic Control*, **19**, 716–723.

Alligood, K. T., J. A. Yorke, and T. D. Sauer, 2000: *Chaos: An Introduction to Dynamical Systems*. Springer, Berlin.

Beckmann, B. and T. A. Buishand, 2002: Statistical downscaling relationships for precipitation in the netherlands and North Germany. *International Journal of Climatology*, **22 (1)**, 15–32, doi:10.1002/joc.718.

Beniston, M., F. Keller, B. Koffi, and S. Goyette, 2003: Estimates of snow accumulation and volume in the Swiss Alps under changing climatic conditions. *Theoretical and Applied Climatology*, **76 (3)**, 125–140, doi:10.1007/s00704-003-0016-5.

Beven, K. J., 2004: *Rainfall-runoff modelling: the primer*. John Wiley and Sons.

Breiling, M. and P. Charamza, 1999: The impact of global warming on winter tourism and skiing: a regionalised model for Austrian snow conditions. *Regional Environmental Change*, **1 (1)**, 4–14, doi:10.1007/s101130050003.

Cao, J. and J. Wang, 2004: Absolute exponential stability of recurrent neural networks with lipschitz-continuous activation functions and time delays. *Neural Networks*, **17 (3)**, 379–390.

Cavazos, T., 2005: Performance of NCEP-NCAR reanalysis variables in statistical downscaling of daily precipitation. *Climate research*, **28**, 95.

Chevallier, F., J. Morcrette, F. Chruy, and N. A. Scott, 2000: Use of a neural-network-based long-wave radiative-transfer scheme in the ECMWF atmospheric model. *Quarterly Journal of the Royal Meteorological Society*, **126 (563)**, 761–776, doi:10.1002/qj.49712656318.

Connor, J. T., R. D. Martin, and L. E. Atlas, 1994: Recurrent neural networks and robust time series prediction. *IEEE Transactions on Neural Networks*, **5 (2)**, 240–254.

Coulibaly, P., Y. B. Dibike, and F. Anctil, 2005: *Downscaling Precipitation and Temperature with Temporal Neural Networks*.

Crane, R. G. and B. C. Hewitson, 2003: Clustering and upscaling of station precipitation records to regional patterns using self-organizing maps (SOMs). *Climate Research*, **25 (2)**, 95–107.

Crawford, T., N. L. Betts, and D. FavisMortlock, 2007: GCM grid-box choice and predictor selection associated with statistical downscaling of daily precipitation over northern ireland. *Climate Research*, **34 (2)**, 145–160, doi:10.3354/cr034145.

Davis, C. and K. Emanuel, 1991: Potential vorticity diagnostics of cyclogenesis. *Monthly Weather Review*, **119**, 1929–1953.

Dawson, C. W. and R. L. Wilby, 2001: Hydrological modelling using artificial neural networks. *Progress in Physical Geography*, 80 –108, doi:10.1177/030913330102500104.

Demuth, H., M. Beale, and M. Hagan, 2005: *Neural Network Toolbox for use with MATLAB, Version 4*. Natik.

D'onofrio, A., J. P. Boulanger, and E. Segura, 2010: CHAC: a weather pattern classification system for regional climate downscaling of daily precipitation. *Climatic Change*, **98 (3)**, 405–427, doi:10.1007/s10584-009-9738-4.

Elsasser, H. and R. Bürki, 2002: Climate change as a threat to tourism in the Alps. *Climate Research*, **20 (3)**, 253–257, doi:10.3354/cr020253.

Elsasser, H. and P. Messerli, 2001: The vulnerability of the snow industry in the Swiss Alps. *Mountain Research and Development*, **21 (4)**, 335–339.

Engelbrecht, A. P. and I. Cloete, 1998: A multilevel nonlinearity study design. *IEEE International World Congress on Computational Intelligence, International Joint Conference on Neural Networks*.

Engelbrecht, A. P. and I. Cloete, 1999: Incremental learing using sensitivity analysis. *Proceedings of IEEE International Joint Conference on Neural Networks*, Vol. 2, 1350–1355.

Falarz, M., 2004: Variability and trends in the duration and depth of snow cover in Poland in the 20th century. *International Journal of Climatology*, **24 (13)**, 1713–1727, doi: 10.1002/joc.1093.

Fan, J. and Q. Yao, 2005: *Nonlinear Time Series: Nonparametric and Parametric Methods*. Springer.

Fealy, R. and J. Sweeney, 2007: Statistical downscaling of precipitation for a selection of sites in Ireland employing a generalised linear modelling approach. *International Journal of Climatology*, **27 (15)**, 2083–2094.

Fowler, H. J., S. Blenkinsop, and C. Tebaldi, 2007: Linking climate change modelling to impacts studies: recent advances in downscaling techniques for hydrological modelling. *International Journal of Climatology*, **27 (12)**, 1547–1578.

Freeman, J. A. and D. M. Skapura, 1991: *Neural Networks - Algorithms, Applications, and Programming Techniques*. Addison-Wesley.

Giannakis, G. and E. Serpedin, 2001: A bibliography on nonlinear system identification. *Signal processing*, **81 (3)**, 533–580.

Goodess, C. M. and J. P. Palutikof, 1998: Development of daily rainfall scenarios for southeast Spain using a circulation-type approach to downscaling. *International Journal of Climatology*, **18 (10)**, 1051–1083, doi:10.1002/(SICI)1097-0088(199808)18.

Grassberger, P. and I. Procaccia, 1983: Measuring the strangeness of strange attractors. *Physica D: Nonlinear Phenomena*, **9 (1-2)**, 189–208, doi:10.1016/0167-2789(83)90298-1.

Hall, D. K., G. A. Riggs, and V. V. Salomonson, 1995: Development of methods for mapping global snow cover using moderate resolution imaging spectroradiometer data. *Remote Sensing of Environment*, **54 (2)**, 127–140, doi:10.1016/0034-4257(95)00137-P.

Hall, D. K., G. A. Riggs, V. V. Salomonson, N. E. DiGirolamo, and K. J. Bayr, 2002: MODIS snow-cover products. *Remote Sensing of Environment*, **83 (1-2)**, 181–194, doi:10.1016/S0034-4257(02)00095-0.

Hamilton, L. C., C. Brown, and B. D. Keim, 2007: Ski areas, weather and climate: time series models for New England case studies. *International Journal of Climatology*, **27 (15)**, 2113–2124.

Hamilton, L. C., D. E. Rohall, B. C. Brown, G. F. Hayward, and B. D. Keim, 2003: Warming winters and new hampshires lost ski areas: an integrated case study. *International Journal of Sociology and Social Policy*, **23 (10)**, 52–73, doi:10.1108/01443330310790309.

Hantel, M., M. Ehrendorfer, and A. Haslinger, 2000: Climate sensitivity of snow cover duration in Austria. *International Journal of Climatology*, **20 (6)**, 615–640, doi:10.1002/(SICI)1097-0088(200005)20.

Hantel, M. and L. Hirtl-Wielke, 2007: Sensitivity of alpine snow cover to european temperature. *International Journal of Climatology*, **27 (10)**, 1265–1275.

Haylock, M. R., G. C. Cawley, C. Harpham, R. L. Wilby, and C. M. Goodness, 2006: Downscaling heavy precipitation over the United Kingdom: A comparison of dynamical and statistical methods and their future scenarios. *International Journal of Climatology*, **26**, 1397–1415.

Hewitson, B. C. and R. G. Crane, 1996: Climate downscaling: techniques and application. *Climate Research*, **07 (2)**, 85–95, doi:10.3354/cr0007085.

Hewitson, B. C. and R. G. Crane, 2002: Self-organizing maps: applications to synoptic climatology. *Climate Research*, **22**, 13–26.

Homma, T. and A. Saltelli, 1996: Importance measures in global sensitivity analysis of nonlinear models. *Reliability Engineering & System Safety*, **52 (1)**, 1–17, doi:10.1016/0951-8320(96)00002-6.

Hong, Y., K. Hsu, S. Sorooshian, and X. Gao, 2004: Precipitation estimation from remotely sensed imagery using an artificial neural network cloud classification system. *Journal of Applied Meteorology*, **43 (12)**, 1834–1853, doi:10.1175/JAM2173.1.

Hsu, K., H. V. Gupta, and S. Sorooshian, 2005: Artificial neural network modeling of the Rainfall-Runoff process. *Water Resources Research*, **31 (10)**, 2517–2530, doi: 199510.1029/95WR01955.

Huang, W., B. Xu, and A. Chan-Hilton, 2004: Forecasting flows in Apalochicola River using neural networks. *Hydrological Processes*, **18**, 2545–2564.

Huth, R., 1999: Statistical downscaling in central Europe: evaluation of methods and potential predictors. *Climate Research*, **13 (2)**, 91–101, doi:10.3354/cr013091.

Huth, R., 2004: Sensitivity of local daily temperature change estimates to the selection of downscaling models and predictors. *Journal of Climate*, **17**, 640–652.

Hyvärinen, A. and E. Oja, 2000: Independent component analysis: algorithms and applications. *Neural Networks*, **13 (4-5)**, 411–430.

IPCC, 2007: *Climate change 2007 : the physical science basis : contribution of Working Group I to the Fourth Assessment Report of the Intergovernmental Panel on Climate Change*. Cambridge University Press, Cambridge ;New York.

Jain, A., K. P. Sudheer, and S. Srinivasulu, 2004: Idendification of physical processes inherent in artificial neural network rainfall runoff models. *Hydrological Processes*, **18**, 571–581.

Kantz, H. and T. Schreiber, 2004: *Nonlinear time series analysis*. Cambridge University Press.

Kasabov, N. K., 1998: *Foundation of Neural Networks, Fuzzy System, and Knowledge Engineering*. The MIT Press, London.

Kecman, V., 2001: *Learning and Soft Computing*. The MIT Press, Cambridge, Massachusetts.

Kirkpatrick, S., C. D. Gelatt, and M. P. Vecchi, 1983: Optimization by simulated annealing. *Science*, **220, 4598**, 671–680.

Bibliography

Klein, A. G., D. K. Hall, and G. A. Riggs, 1998: Improving snow cover mapping in forests through the use of a canopy reflectance model. *Hydrological Processes*, **12 (10-11)**, 1723–1744, doi:10.1002/(SICI)1099-1085(199808/09)12.

Knutti, R., T. F. Stocker, F. Joos, and G. Plattner, 2003: Probabilistic climate change projections using neural networks. *Climate Dynamics*, **21 (3-4)**, 257–272, doi:10.1007/s00382-003-0345-1.

Kohonen, T., 1982: Self-organized formation of topologically correct feature maps. *Biological Cybernetics*, **43 (1)**, 59–69, doi:10.1007/BF00337288.

Krasnopolsky, V. M., M. S. Fox-Rabinovitz, and D. V. Chalikov, 2005: New approach to calculation of atmospheric model physics: Accurate and fast neural network emulation of longwave radiation in a climate model. *Monthly Weather Review*, **133 (5)**, 1370–1383, doi:10.1175/MWR2923.1.

Kwak, N. and C. Choi, 2002: Input feature selection for classification problems. *IEEE Transactions on Neural Networks*, **13 (1)**, 143–159.

Laternser, M. and M. Schneebeli, 2003: Long-term snow climate trends of the Swiss Alps (1931-99). *International Journal of Climatology*, **23 (7)**, 733–750, doi:10.1002/joc.912.

Lin, T., B. Horne, P. Tino, and C. Giles, 1996: Learning long-term dependencies in NARX recurrent neural networks. *Neural Networks, IEEE Transactions on*, **7 (6)**, 1329–1338, doi: 10.1109/72.548162.

Lynch, A., P. Uotila, and j. J Cassano, 2006: Changes in synoptic weather patterns in the polar regions in the twentieth and twenty-first centuries, part 2: Antarctic. *International Journal of Climatology*, **26**, 1181–1199.

Maier, H. R. and G. C. Dandy, 1998: The effect of internal parameters and geometry on the performance of backpropagation neural networks: an empirical study. *Environmental Modelling and Software*, **13**, 193–209.

Maier, H. R. and G. C. Dandy, 2000: Neural networks for the prediction and forecasting of water resources variables: a review of modelling issues and applications. *Environmental Modelling and Software*, **15**, 101–123.

Maraun, D., et al., 2010: Precipitation downscaling under climate change. recent developments to bridge the gap between dynamical models and the end user. *Rev. Geophys.*, doi: 10.1029/2009RG000314.

Metropolis, N., A. W. Rosenbluth, M. N. Rosenbluth, A. H. Teller, and E. Teller, 1953: Equation of state calculations by fast computing machines. *The Journal of Chemical Physics*, **21 (6)**, 1087, doi:10.1063/1.1699114.

Möller, M., C. Schneider, and R. Kilian, 2007: Glacier change and climate forcing in recent decades at Gran Campo Nevado, southernmost Patagonia. *Annals of Glaciology*, **46**, 136–144, doi:10.3189/172756407782871530.

Moreau, A., O. Teytaud, and J. P. Bertoglio, 2006: Optimal estimation for large-eddy simulation of turbulence and application to the analysis of subgrid models. *Physics of Fluids*, **18 (10)**, 105–101, doi:10.1063/1.2357974.

Murphy, J., 2000: Predictions of climate change over europe using statistical and dynamical downscaling techniques. *International Journal of Climatology*, **20**, 489–501.

Nash, J. and J. Sutcliffe, 1970: River flow forecasting through conceptual models part i - a discussion of principles. *Journal of Hydrology*, **10 (3)**, 282–290, doi:10.1016/0022-1694(70)90255-6.

Newton, S. I., 1999: *The Principia: Mathematical Principles of Natural Philosophy*. 1st ed., University of California Press, translation by Bernhard I. Cohen and Anne Whitman and Julia Budenz.

Orlowsky, B., F. W. Gerstengarbe, and P. C. Werner, 2007: A resampling scheme for regional climate simulations and its performance compared to a dynamical RCM. *Theoretical and Applied Climatology*.

Philipp, A., P. M. Della-Marta, J. Jacobeit, D. R. Fereday, P. D. Jones, A. Moberg, and H. Wanner, 2007: Long-term variability of daily north Atlantic-European pressure patterns since 1850 classified by simulated annealing clustering. *Journal of Climate*, **20 (16)**, 4065–4095.

Poddig, T. and S. Sidorovitch, 2001: Künstliche neuronale netze: Überblick, einsatzmöglichkeiten und anwendungsprobleme. *Handbuch Data Mining im Marketing (Hrsg. H. Hippner, U. Kösters, M. Meyer und K. Wilde)*, 363–402.

Ratto, M. and A. Saltelli, 2001: *Gluewin users manual*.

(REKLIP), T. A. R., 1995: *Klimaatlas Oberrhein Mitte-Süd*. Offenbach.

Riad, S., J. Mania, L. Bouchaou, and Y. Najjar, 2004: Predicting catchment flow in a semi-arid region via an artificial neural network technique. *Hydrological Processes*, **18**, 2387–2393.

Rissanen, J., 1978: Modeling by shortest data description. *Automatica*, **14 (5)**, 465–471, doi:10.1016/0005-1098(78)90005-5.

Roebber, P. J., M. R. Butt, S. J. Reinke, and T. J. Grafenauer, 2007: Real-Time forecasting of snowfall using a neural network. *Weather and Forecasting*, **22 (3)**, 676–684.

Rosenstein, M. T., J. J. Collins, and C. J. D. Luca, 1993: A practical method for calculating largest Lyapunov exponents from small data sets. *Physica D: Nonlinear Phenomena*, **65 (1-2)**, 117–134, doi:10.1016/0167-2789(93)90009-P.

Rossa, A. M., H. Wernli, and H. C. Davies, 2000: Growth and decay of an Extra-Tropical cyclone's PV-Tower. *Meteorology and Atmospheric Physics*, **73 (3-4)**, 139–156, doi: 10.1007/s007030050070.

Salomonson, V. V. and I. Appel, 2004: Estimating fractional snow cover from MODIS using the normalized difference snow index. *Remote Sensing of Environment*, **89 (3)**, 351–360, doi: 10.1016/j.rse.2003.10.016.

Saltelli, A., M. Ratto, S. Tarantola, and F. Campolongo, 2006: Sensitivity analysis practices: Strategies for model-based inference. *Reliability Engineering & System Safety*, **91 (10-11)**, 1109–1125, doi:10.1016/j.ress.2005.11.014.

Saltelli, A., S. Tarantola, and F. Campolongo, 2000: Sensitivity analysis as an ingredient of modeling. *Statistical Science*, **15 (4)**, 877–895.

Saltelli, A., S. Tarantola, F. Campolongo, and M. Ratto, 2004: *Sensitivity Analysis in Practice: A guide to assessing scientific models*. John Wiley and Sons, Chichester.

Saltelli, A., S. Tarantola, and K. P. Chan, 1999: A quantitative Model-Independent method for global sensitivity analysis of model output. *Technometrics*, **41 (1)**, 39–56.

Sarghini, F., G. de Felice, and S. Santini, 2003: Neural networks based subgrid scale modeling in large eddy simulations. *Computers & Fluids*, **32 (1)**, 97–108, doi:10.1016/S0045-7930(01)00098-6.

Sauter, T., C. Schneider, R. Kilian, and M. Moritz, 2009: Simulation and analysis of runoff from a partly glaciated meso-scale catchment area in Patagonia using an artificial neural network. *Hydrological Processes*, **23 (7)**, 1019–1030.

Schneider, C., M. Glaser, R. Kilian, A. Santana, N. Butorovic, and G. Casassa, 2003: Weather observations across the southern andes at 53S. *Physical Geography*, **24**, 97–119.

Schneider, C., G. Ketzler, C. Maas, M. Buttstädt, M. Möller, T. Sauter, and B. Weitzenkamp, 2007a: Statistische analyse von liftbetriebszeiten in abhängigkeit von klimaelementen am beispiel des Südschwarzwaldes. *Unveröffentlichtes Gutachen für die Stiftung Sicherheit im Skisport des Deutschen Skiverbandes*.

Schneider, C., R. Kilian, and M. Glaser, 2007b: Energy balance in the ablation zone during the summer season at the Gran Campo Nevado Ice Cap in the Southern Andes. *Global and Planetary Change*, **59 (1-4)**, 175–188, doi:10.1016/j.gloplacha.2006.11.033.

Schneider, C., H. Saurer, and J. Schönbein, 2005: Schneesport ohne schnee? *Praxis Geographie*, **35 (5)**, 18–23.

Schneider, C. and J. Schönbein, 2006: Klimatologische analyse der schneesicherheit und beschneibarkeit von wintersportgebieten in deutschen mittelgebirgen. *Schriftenreihe Natursport und Ökologie*, **111**.

Schönbein, J. and C. Schneider, 2005: Zur klimatologie der winterlichen schneedecke deutscher mittelgebirge. *Geoöko*, **26**, 197–216.

Schoof, J. T. and S. C. Pryor, 2001: Downscaling temperature and precipitation: A comparison of regression-based methods and artificial neural networks. *International Journal of Climatology*, **21**, 773–790.

Schreiber, T. and A. Schmitz, 1996: Improved surrogate data for nonlinearity tests. *Physical Review Letters*, **77 (4)**, 635, doi:10.1103/PhysRevLett.77.635.

Schreiber, T. and A. Schmitz, 2000: Surrogate time series. *Physica D: Nonlinear Phenomena*, **142 (3-4)**, 346–382.

Scott, D., J. Dawson, and B. Jones, 2008: Climate change vulnerability of the US northeast winter recreation- tourism sector. *Mitigation and Adaptation Strategies for Global Change*, **13 (5)**, 577–596, doi:10.1007/s11027-007-9136-z.

Scott, D., G. McBoyle, and A. Minogue, 2007: Climate change and Quebec's ski industry. *Global Environmental Change*, **17 (2)**, 181–190, doi:10.1016/j.gloenvcha.2006.05.004.

Shamseldin, A. Y., 1997: Application of a neural network technique to rainfall-runoff modelling. *Journal of Hydrology*, **199**, 272–294.

Siegelmann, H. T., B. G. Horne, and C. L. Giles, 1997: Computational capabilities of recurrent NARX neural networks. *Systems, Man and Cybernetics, Part B, IEEE Transactions on*, **27 (2)**, 208–215.

Simpson, J. J. and T. J. McIntire, 2001: A recurrent neural network classifier for improved retrievals of areal extent of snow cover. *IEEE Transactions on Geoscience and Remote Sensing*, **39 (10)**, 2135–2147.

Sobol, I. M., 2001: Global sensitivity indices for nonlinear mathematical models and their Monte Carlo estimates. *Mathematics and Computers in Simulation*, **55 (1-3)**, 271–280, doi: 10.1016/S0378-4754(00)00270-6.

Solomatine, D. P., M. Maskey, and D. L. Shrestha, 2008: Instance-based learning compared to other data-driven methods in hydrological forecasting. *Hydrological Processes*, **22 (2)**, 275–287, doi:10.1002/hyp.6592.

Storch, H. and F. W. Zwiers, 1999: *Statistical Analysis in Climate Research*. Cambridge University Press, Cambridge.

Strogatz, S. H., 2001: *Nonlinear Dynamics And Chaos: With Applications To Physics, Biology, Chemistry, And Engineering*. 1st ed., Westview Press.

Sudheer, K. P., A. K. Gosain, and K. S. Ramasastri, 2002: A data-driven algorithm for constructing artificial neural network rainfall-runoff models. *Hydrological Processes*, **16**, 1325–1330.

Sudheer, K. P. and A. Jain, 2004: Explaining the internal behaviour of artificial neural network river flow models. *Hydrological Processes*, **18**, 833–844.

Supharatid, S., 2003: Application fo a neural network model in establishing a stage-discharge relationship for a tidal river. *Hydrological Processes*, **17**, 3085–3099.

Taha, I. A. and J. Ghosh, 1999: Symbilic interpretation of artificial neural networks. *IEEE Transactions on knowledge and data enineering*, **11 (3)**, 448–463.

Tappeiner, U., G. Tappeiner, J. Aschenwald, E. Tasser, and B. Ostendorf, 2001: GIS-based modelling of spatial pattern of snow cover duration in an alpine area. *Ecological Modelling*, **138 (1-3)**, 265–275.

Theiler, J., S. Eubank, A. Longtin, B. Galdrikian, and J. D. Farmer, 1992: Testing for nonlinearity in time series: the method of surrogate data. *Physica D: Nonlinear Phenomena*, **58 (1-4)**, 77–94, doi:10.1016/0167-2789(92)90102-S.

Tolika, K., P. Maheras, M. Vafiadis, H. A. Flocas, and A. Arseni-Papadimitriou, 2007: Simulation of seasonal precipitation and raindays over greece: a statistical downscaling technique based on artificial neural networks (ANNs). *International Journal of Climatology*, **27**, 861–881.

Tsoukalas, L. H., R. E. Uhrig, and L. A. Zadeh, 1997: *Fuzzy and Neural Approaches in Engineering*. John Wiley & Sons, Chichester.

Ulbrich, U., G. Leckebusch, and J. Pinto, 2009: Extra-tropical cyclones in the present and future climate: a review. *Theoretical and Applied Climatology*, **96 (1)**, 117–131, doi: 10.1007/s00704-008-0083-8.

Ulbrich, U., J. G. Pinto, H. Kupfer, G. C. Leckebusch, T. Spangehl, and M. Reyers, 2008: Changing northern hemisphere storm tracks in an ensemble of IPCC climate change simulations. *Journal of Climate*, **21 (8)**, 1669–1679, doi:10.1175/2007JCLI1992.1.

Uvo, C. B., U. Tlle, and R. Berndtsson, 2000: Forecasting discharge in Amazonia using artificial neural network. *International Journal of Climatology*, **20**, 1495–1507.

Van der Smagt, P. P. and G. Hirzinger, 1996: Solving the Ill-Conditioning in neural network learning. *Neural Networks: Tricks of the Trade*, 193–206.

Venema, V., S. Bachner, H. W. Rust, and C. Simmer, 2006a: Statistical characteristics of surrogate data based on geophysical measurements. *Nonlinear Processes in Geophysics*, **13**, 449–466.

Venema, V., et al., 2006b: Surrogate cloud fields generated with the iterative amplitude adapted Fourier transform algorithm. *Tellus A*, **58 (1)**, 104–120.

Wang, X. L., V. R. Swail, and F. W. Zwiers, 2006: Climatology and changes of extratropical cyclone activity: Comparison of ERA-40 with NCEP-NCAR reanalysis for 1958-2001. *Journal of Climate*, **19**, 3145–3166.

Weischet, W. and W. Endlicher, 2000: *Regionale Klimatologie, Teil 2, Die alte Welt*. Gebr. Borntraeger, Stuttgart.

Wilby, R. and T. Wigley, 2000: Precipitation predictors for downscaling: observed and general circulation model relationships. *International Journal of Climatology*, **20 (6)**, 641–661, doi:10.1002/(SICI)1097-0088(200005)20.

Wilby, R. L., R. J. Abrahart, and C. W. Dawson, 2003: Detectin of conceptual model rainfall-runoff processes inside an artificial neural network. *Hydrological Science Journal*, **48 (2)**, 163–181.

Wilby, R. L. and T. M. Wigley, 1997: Downscaling general circulation model output: a review of methods and limitations. *Progress in Physical Geography*, **21 (4)**, 530–548.

Wilby, R. L., T. M. L. Wigley, D. Conway, P. D. Jones, B. C. Hewitson, J. Main, and D. S. Wilks, 1998: Statistical downscaling of general circulation model output: A comparison of methods. *Water Resources Research*, **34 (11)**, 2995–3008, doi:199810.1029/98WR02577.

Wilks, D. S., 2006: *Statistical Methods in the Atmospheric Sciences*. Elsevier, San Diego, CA.

Zell, A., 2003: *Simulation Neuronaler Netze*. Oldenbourg Wissenschaftsverlag.

Acknowledgements

First of all I wish to thank my supervisor Prof. Dr. Christoph Schneider who has given me the opportunity to explore my research interests in the last four years. He also gave me a number of special opportunities to participate in various field campaigns, such Patagonia, Svalbard, Tibet and the Black Forest. Without his guidance and confidence it would have been much harder. During the period of the GCN project I closely worked together with Björn Weizenkamp, albeit without spending some good times on field campaigns and after work sessions. This acknowledgements would be rather incomplete without expressing my gratitude to him. Special thanks go to Dr. Victor Venema, for the intensive cooperation throughout the last two years. I very much appreciate his advices and the endless but fruitful discussions. Further, I thank my room mate and friend Oliver Käsmacher for the pleasant time in Aachen and the enriching talks on everything and anything. I am equally grateful to Dr. Gunnar Ketzler, who made my new start in Aachen so much easier by giving me some shelter and a comprehensive introduction to local history. I also very much appreciate the advice and inspiring discussions with Priv.-Doz. Dr. Wolgang Römer. I thank Eva Huintjes, Marco Möller, Dr. Katja Petzholdt, Mareike Buttstädt, Hendrik Merbitz, Timo Sachsen, Elke Bruer and all the student assistent for the nice time we had. For many nice evenings and discussions, I also would like to thank my geomorphology colleagues René Löhrer and Georg Stauch.

My very personal thanks and wishes are dedicated to my wife Jimena Zurieta.

i want morebooks!

Buy your books fast and straightforward online - at one of world's fastest growing online book stores! Environmentally sound due to Print-on-Demand technologies.

Buy your books online at
www.get-morebooks.com

Kaufen Sie Ihre Bücher schnell und unkompliziert online – auf einer der am schnellsten wachsenden Buchhandelsplattformen weltweit! Dank Print-On-Demand umwelt- und ressourcenschonend produziert.

Bücher schneller online kaufen
www.morebooks.de

VDM Verlagsservicegesellschaft mbH
Heinrich-Böcking-Str. 6-8
D - 66121 Saarbrücken

Telefon: +49 681 3720 174
Telefax: +49 681 3720 1749

info@vdm-vsg.de
www.vdm-vsg.de

Printed by Books on Demand GmbH, Norderstedt / Germany